SpringerBriefs in Computer Science

SpringerBriefs present concise summaries of cutting-edge research and practical applications across a wide spectrum of fields. Featuring compact volumes of 50 to 125 pages, the series covers a range of content from professional to academic.

Typical topics might include:

- A timely report of state-of-the art analytical techniques
- A bridge between new research results, as published in journal articles, and a contextual literature review
- A snapshot of a hot or emerging topic
- An in-depth case study or clinical example
- A presentation of core concepts that students must understand in order to make independent contributions

Briefs allow authors to present their ideas and readers to absorb them with minimal time investment. Briefs will be published as part of Springer's eBook collection, with millions of users worldwide. In addition, Briefs will be available for individual print and electronic purchase. Briefs are characterized by fast, global electronic dissemination, standard publishing contracts, easy-to-use manuscript preparation and formatting guidelines, and expedited production schedules. We aim for publication 8–12 weeks after acceptance. Both solicited and unsolicited manuscripts are considered for publication in this series.

**Indexing: This series is indexed in Scopus, Ei-Compendex, and zbMATH **

Janos J. Sarbo

Inevitable Knowledge

 Springer

Janos J. Sarbo
Utrecht, The Netherlands

ISSN 2191-5768   ISSN 2191-5776   (electronic)
SpringerBriefs in Computer Science
ISBN 978-3-031-73460-1   ISBN 978-3-031-73461-8   (eBook)
https://doi.org/10.1007/978-3-031-73461-8

This Springer imprint is published by the registered company Springer Nature Switzerland AG
The registered company address is: Gewerbestrasse 11, 6330 Cham, Switzerland

If disposing of this product, please recycle the paper.

If you keep on knocking long enough at the same door it always opens in the end. Or else there is a chink in another door nearby that you hadn't noticed; and that's even better.
[Michel Tournier: The Ogre. Translated by Barbara Bray; Johns Hopkins University Press, 1997]

Prologue

Technological revolutions change the world in an unprecedented way. Besides replacing old technology with new one, they rewrite the material conditions of human existence and reshape culture. In addition to the bare possibility of new technology, the driving force behind this metamorphosis is a novel view of the world. An example is engineering, as a paradigm for solving all problems, characterizing the transition from manual production methods to machine production during the Industrial Revolution in the nineteenth century.

The perspective that characterizes the current technological revolution is information, not just as data but as a mode of organization and functioning. All things and properties, therefore all knowledge, are viewed from this angle. The information revolution promises to transform the world into an organization providing meaningful solutions to problems.

Communication theory conceives information as the probabilities of the occurrences of symbols in their respective context (Shannon, 1948). A typical example of a symbol is a letter occurring in a message. The complexity of information (e.g., storage size) is expressed by the number of bits necessary for encoding all possible symbol occurrences. With the increasing use of computer technology, the focus has shifted from information as data to information as knowledge, heralding a new age in which the complexity of knowledge a program can deal with is more important than the storage size required.

The information technology revolution has been underway for some time. Yet, we are still far from seeing the world as a "meaningful" computational system. In particular, this goal can only be achieved by respecting the properties of interpretation, that is, meaningful processing, the importance of which has yet to be sufficiently realized.

Meaningful information processing is rooted in a wide range of disciplines. The concepts of meaning and interpretation fall in philosophy, the conditions of human processing in cognitive theory, and the properties of information as a formal notion lie at the heart of computer science. Research at the interface of these disciplines is concerned with the types of meaningful expression, the properties of human processing, and how those types and properties can be captured in an algorithm.

An account of such interdisciplinary research is the goal set for this book. We borrow a theory of interpretation from a philosophical framework of signs and a theory of cognitive processing from a study of the human visual system and combine them in an information-processing model. An unexpected outcome of this undertaking is that the ground for an important phenomenon related to thinking, namely consciousness, could be architectural features of brain processing.

Traditionally, interpretation refers to human processing. This book aims to show that this concept can be applied more broadly. The connection between structural properties of the model of meaningful processing and fundamental natural phenomena, such as spherical waves, suggests that all of nature must be capable of interpretation. Knowledge must be inevitable.

Acknowledgments

I am grateful to E. Taborsky and J.I. Farkas for their valuable comments and to R. Cozijn for his continuing support.

Contents

Contents

Chapter 1
Introduction

Abstract Human processing deals with phenomena that appear in our perception. Meaning arises through interpretation; knowledge can be seen as learned information. Phenomena occur as a stimulus and interact with the brain/mind as an interpreting system, transforming the input into an environmentally significant response. Since this view of information processing applies to all modalities, we assume the input stimulus functions as a sign. Given the general nature of signs, we argue that human interpretation uses one type of process.

1.1 The Curious Nature of Knowledge

There is something curious about knowledge. We have it and do not have it at the same time. Although we have vast knowledge about the world and can use it effortlessly to solve problems, we cannot give an exhaustive answer if asked to tell everything we know. What could explain this feature of our brain/mind?[1] To answer this question, we delve into the properties of human processing. At the outset, we ask: What is its source, and how do we obtain knowledge?

As for its source, we assume that knowledge arises from phenomena, i.e., what appears in the world. However, this answer is too superficial. It ignores that knowledge emerges from perception, that is, experience. Therefore, the correct answer must be that the source of knowledge is phenomena that occur to us as stimuli. While accurate, it does not illuminate an essential stimuli characteristic: they usually stand for something other than themselves and thus function as signs. For instance, you hear train noise, but that indicates danger. Only through interpreting stimuli, which themselves are phenomena, can we find out what they signify. From this, we conclude that knowledge must arise from phenomena processed as signs. The process itself can be explained as follows. The external stimulus triggers memory, and the arising information can be used to reveal the reason for the

---

[1] The brain is usually seen as a physical organ where the mind resides. Mind refers to thought processes and reason. Interpretation and meaning are assumed to be a product of the mind. We believe that the brain and the mind are not different.

© The Author(s), under exclusive license to Springer Nature Switzerland AG 2025
J. J. Sarbo, *Inevitable Knowledge*, SpringerBriefs in Computer Science,
https://doi.org/10.1007/978-3-031-73461-8_1

perceived data. To this end, interpretation transforms the input into a response by establishing a link between the stimulus and memory information. To understand the properties of this transformation, we must look at the conditions of information processing by the brain as an *interpreting system*. Since interpretation is related to data, we also must pay attention to memory organization, including the representation of stored data. Both issues are complex. Mastering them requires a thorough analysis of information processing and cognitive activity properties.

Two comments. First, the choice of interpreting system, rather than interpreter or agent, favored by Artificial Intelligence (AI), is not accidental. The concept of interpretation can be applied broadly, not just to human processing. Second, we use knowledge and information interchangeably, assuming that knowledge can be defined as learned information.

1.1.1   Historical Attempts

All disciplines have an ideal or a holy grail. For AI, this must be a machine that can think human-like. The thought that human thinking can be subject to engineering is old and has fascinated humanity for centuries. We skip those mythological ideas, which can be interesting from a historical point of view but not helpful for our goals, and concentrate on a couple of more 'realistic' attempts in the past.

The first is the mechanical chess player, "The Turk," invented by Wolfgang von Kempelen. His machine, which won against many, including politicians such as Napoleon and Franklin, was hiding a master chess player dwarf. The deception was complete. Many believed the machine to be human-like because it behaved like a human, which it partly was.

Nevertheless, what makes von Kempelen's creation unique is the amount of credit it received and why. Van Kempelen was a successful inventor of his time, which must have fueled the belief that his machine could think. The appearance of his machine also played a role. We tend to humanize animals in our environment, as well as non-animate objects. Had van Kempelen's machine not looked human-like, it probably would have gotten less credit. However, a human-like appearance alone is not enough. It is also required that moves by the machine can be conceived as strategical chess playing. Over time, interest in the chess machine declined, probably because it did not meet all expectations. While "The Turk" still occasionally gave performances, it was eventually relegated to the corners of the museum and forgotten about until 5 July 1854, when a fire that started at the National Theater in Philadelphia reached the museum and destroyed von Kempelen's mechanical chess player.[2]

The second attempt we mention is the concept of a homunculus or the assumption of a putative process behind the brain's decision-making. Chomsky's Universal Grammar, a theory in linguistics, is rooted in the same idea. However, the notion

---

[2]https://en.wikipedia.org/wiki/Mechanical_Turk

of a homunculus needs to be revised, as it suffers from the danger of infinite regression. If the homunculus governs our thinking, what governs the decisions made by the homunculus?

A dubious problem with the concept of a homunculus, which is more critical to our goals, is the assumption that knowledge can be captured in facts. Chomsky's theory is a good illustration of this idea. The fundamental postulate of Universal Grammar is that a specific set of structural rules is innate to humans. Advocates of his theory claim that natural language has universal properties, but this conjecture has yet to be firmly established. Nevertheless, Chomsky's theory significantly impacted computational language modeling in the 1970s (Cook and Newson 2007). The idea that natural language can be specified in terms of formal grammar was found attractive, partly because of the clarity of definition and the ability to prove a few properties automatically. It soon became apparent that language definitions can quickly require thousands of formal rules, making maintenance extraordinarily labor-intensive and unfeasible. The blind faith in the hypothesis in the 70 s and today that natural language can be captured in factual rules is astonishing.

The moral of von Kempelen's chess machine is that the operations of any human processing model must appear logical in our experience. The lesson learned from the homunculus, especially from its regression problem, is that formal models of human data processing may not be different from the process of interpretation; their structure must be identical, in technical terms, isomorphic. It is not our factual knowledge of the language that matters but the nature of human processing.

What logic should be respected exactly (see von Kempelen's machine), needs to be clarified. Human reasoning is unlikely to be one of the formal theories of logic. For if it were, it would be easily predictable, which it is not. It also needs to be clarified what conditions must be met by a formal specification to respect human processing and thus be meaningful and how interpretation can be modeled as a process, which brings us to the fundamental question of this book: What is involved in human processing?

1.1.2   Contemporary Attempts

The advent of fast processors and large memory in the 1990s gave an unprecedented new impetus to reach the complexity of human intelligence. One of the first results in 1996, was the chess program Deep Blue (Feng-Hsiung 2004) defeating reigning world champion Kasparov. Another achievement, in 2011, was the Watson question-answering system, competing on *Jeopardy!* against human players.[3] A more recent example is the self-learning system Alpha-Go,[4] defeating 9-dan master Lee Sedol in 2016 (the strategic board game Go is computationally much more

---

[3] IBM's Watson failed in healthcare practice. See https://nl.wikipedia.org/wiki/DeepMind

[4] https://en.wikipedia.org/wiki/AlphaGo

complex than chess),[5] mastering the game of Stratego (Perolat et al. 2022) and applications in text processing, art creation,[6] etc., in 2022.

The skill of these programs amazed everyone. Indeed, if those programs can handle such complex problems, why can't they take all the tasks humans can? And yet, and this is equally astonishing, no one came forward and claimed that the computer could outperform the brain, despite the common belief that superiority by the computer is not unthinkable. The argument that computers are just machines is not helpful. What could explain our distrust of the computer?

Deep in our hearts, we feel the computer works differently from natural systems, especially our brain/mind. The hand-waving answer that brain processing must involve a qualitative change does not help either. Its conditions are still a mystery. We suggest an alternative explanation without denying the possibility of such a change. According to this, their difference and thus the distrust in the computer could result from the different ways the computer and the brain process information. Not how it is realized (this point is obvious), but the kinds of operations performed. The computer can run any program, unlike the brain, which, on the premise of this book, performs one procedure. We trust our brains, but we can have less faith in the results of programs because of their formal nature and lack of transparency about how they come about.[7] Our distrust must be related to the earlier question raised by this book, namely the nature of interpretation and phenomena as the source of knowledge. That said, let us see what the brain's procedure could be.

1.2   What Is Needed to Recognize Phenomena?

According to our assumption, knowledge arises through observation. In practice, this means that external phenomena appear as stimuli in sensory perception. We can only recognize stimuli we are already familiar with (the morphology of 're-cognition' also indicates this meaning). How do we deal with unknown phenomena? In the extreme case when information is limited to sensory data, we can cognize the input as it is, so only as an existence.

External stimuli[8] appear in our perception as properties such as color, temperature, mass, etc. A more comprehensive concept is *quality*, which also means differentiation in addition to a property. Concerning knowledge representation, the latter is essential, which explains why we will continue to use this notion. An example of a quality is the activity of receptors in the retina during visual perception.

---

[5] https://arstechnica.com/information-technology/2022/11/new-go-playing-trick-defeats-world-class-go-ai-but-loses-to-human-amateurs/

[6] ChatGTP, Dall-E. See https://nl.wikipedia.org/wiki/ChatGPT

[7] Because it can provide false information, we cannot fully trust ChatGPT. See Kamalloo et al. (2023).

[8] External, to the brain as an interpreting system.

The question of how this activity can function as differentiation is for later, in Chap. 3. Regarding information organization, receptor activation can refer to a single receptor or a set of receptors. Since all stimuli have an analogous twofold meaning, we assume that quality can identically denote a single quality and a set of qualities.

Mediated by sensory signals, the stimulus activates memory information about phenomena observed in the past. In principle, only data similar to the input will answer. Depending on the degree to which the stimulus matches stored information, memory may respond in agreement or possibility sense. The first means that we are familiar with this stimulus; the second indicates that we know only a subset of its qualities, i.e., we know it approximately. In both cases, the memory response may consist of a collection of data. Memory information that is not responsive to the stimulus is not considered in input processing.[9]

Input recognition is successful only if the arising memory response allows for an explanation of the stimulus as a whole. The two sorts of memory responses facilitate two modes of information processing. The first is when the whole collection of input qualities matches memory in the sense of agreement, i.e., above a threshold. In this case, the input can be recognized immediately, and its meaning as a sensuous impression is obvious. We call this *matching mode* processing. The second is when matching in the sense of agreement is only available for a subset of qualities (remember that some familiarity with the input is necessary for recognition); the remaining subset matches memory only in the sense of possibility or below threshold. Since the stimulus is a phenomenon, that is, a coherent set of qualities, the responding memory information must also possess the property of coherency.[10] This similarity implies that agreement- and possibility-sense memory information must be complementary. How does this affect input processing?

When matching mode processing is unavailable, possibility-sense memory response can explain the difference between the stimulus and memory response in the sense of agreement. In other words, its information can reveal how the stimulus, as it is now, differs from what it was, according to stored knowledge. If the difference can be explained this way, the entire input can be recognized, just like in matching mode processing. The above way of input processing can be cumbersome. It is like tiling the qualities of the stimulus with memory information used as patterns, just as in a jigsaw puzzle. The set of complementary data may be significant, and finding an explanation for the whole stimulus may require its elements to be selected based on some criterion, for example, visited individually. We call this *analysis mode* processing. A remarkable consequence of matching as an elementary

---

[9]Response information may depend on the temporal conditions of memory activation, including learning ability. Suppose you observe a phenomenon in a given period. In that case, you will likely use the same memory information to process a later occurrence of the same or similar phenomenon. The conditions for learning, temporary or permanent, are beyond our focus.

[10]Remember that memory information is about phenomena observed in the past.

Fig. 1.1 A level crossing
with a stop-board

operation of input recognition, which applies to both processing modes, is that stimulus and memory information must be represented identically using qualities.

Take the train sound at a railway crossing as an example of a stimulus. See Fig. 1.1. We perceive the sound only as a quality but recognize it as a warning of danger. What does it take to come to this conclusion? First, we recall that sound signals can be analyzed into periodic waves of different frequencies. Analogously, our hearing mechanism sends messages to the brain that differ depending on the frequency content of the sound signal (Hewlett and Beck 2013). Suppose we have knowledge matching the frequency constituents of the perceived sound above a threshold value. In that case, we can immediately recognize our input as a signal. If this fails, but we can recognize a subset of the input periodic waves as train noise (in an agreement sense) and identify the remaining frequency constituents as a rumbling sound, we can still conclude that the noise is a signal of danger (in the sense of possibility). With more knowledge, we can even deduce that the input sound is a warning of the danger of an approaching train, which may also be our best guess in the presence of the visual qualities of a level crossing.

1.2.1   Static Versus Dynamic

The relation between stimulus and response (or meaning) can be thought of as static and dynamic, which are qualitatively different. Conceptually, the first sees meaning as a value defined mechanically by a relation. According to the second, meaning arises through transformation, producing an effect. Such a relationship can be broken down into components, but in doing so, its meaning disappears. While the components individually can be transformational, their relationship will be static. As a result, the entire transformation loses its dynamic character. This property also applies to information processing. It is quite possible to analyze a dynamic process into events and to talk about their properties, but only in the sense that those events exist within that process, so never separately.

So far, we have seen two cases of transformation: from stimulus into memory response, which represents the stimulus in itself, and from that memory response into a relation with complementary information, which expresses the stimulus in context. In Chap. 3, we will argue that there is a third kind of transformation, from the stimulus in context into its meaning, such as a reaction.

How is a dynamic transformation different from a mechanical translation? In essence, transformation means change. A response to a stimulus originates in and from the brain and physically alters it. This ability is not available for the computer. Computer calculations only change memory contents but always leave the computer's structure as it is.[11] The use of neurotransmitters by memory demonstrates that interpretation is more than mechanical translation. The fact that neurochemicals are used and thus must be re-produced indicates the transformational nature of human processing. How neurotransmitters relate to meaningful processing still needs to be clarified. Surprisingly, there is evidence of a transformation in the brain that can be directly linked to interpretation, but more on this later, in Chap. 3. For completeness, we mention that computer calculations also have an effect due to electrical dissipation, namely heat, but that is less varied than the transformations characteristic of brain processing.

1.2.2   Phenomena Are Signs

Interpretation is not a matter of choice. It is indispensable for survival. We must know if the external stimulus can be interesting for any reason, and if so, what can be an environmentally significant response. According to recent neuro-physiological research, human processing is driven by expected information gain and, thus, curiosity-driven (Van Lieshout et al. 2020).

An example of a stimulus, quality, and response—in this order—is the sound of a fast-moving train, the arising vibration in the eardrum, and the thought that a train is coming, so we should wait until it has passed. Another example is a stopboard in the perception of a car driver. See Fig. 1.1. The visual signal of the stopboard is a stimulus; the activity of the retina, triggered by the shape, blue color, and the text "WACHT" (Dutch for "wait"), is a quality; stopping the car is a response by the driver.

---

[11] The computer translates the input, a number, into output, another number. In this sense, computer operations do not change the computer architecture and are, therefore, static.

When the train sound and the visual signal of the stop board convey information about danger and the need to stop the car, they represent something other than themselves and thus act as a *sign*. Technically, what a sign stands for is called the sign's *object*; the effect of a sign is referred to as the sign's *meaning*. From the assumption that signs appear to us, and in general to an interpreting system, as a quality, it follows that all signs involve experience. Here, we focus on signs as qualities and how they are processed into meaning. The philosophical concept of a sign, which mainly concentrates on meaning and, to a lesser extent, on the underlying process, is beyond our horizon.

The three concepts –sign, object, and meaning– are related, and interpretation aims to establish this relation. In this course, the stimulus, as a sign, is transformed from quality into an effect through mediation by the object. Here, the sign, perceived as quality, is regarded as data, the object as access to stored knowledge (triggered by the sign), and the effect of the sign as meaning or response (indicated by the sign and the object). All three concepts involve a transformation, hence a change.

A characteristic of the transformation from stimulus to response is the possibility of the result, such as a thought, to function as a quality in a subsequent interpretation process. In this way, complex stimuli (and phenomena) can be interpreted through a chain of (embedded) processes. For instance, the perceived sound can be recognized first as train noise, then as a warning of danger, and finally as a thought that prompts us to wait for the train to pass.

That looks nice, but is the assumption of such a chain consistent with the conditions set for interpretation so far? How can such a chain exist if the source of knowledge is qualities that appear in sensory perception and meaning arises in the brain, not in the senses?

The assumption that a response can become a quality implies that the brain must be able to modify its input, not the sensory input but the qualities offered for interpretation. An example is when a subset of input qualities is recognized as train noise, and this thought as a new input quality, along with other input qualities, the brain eventually transforms into a trigger for a motor action (e.g., wait for the train to pass). If the response indicates new information (cf. quality), hence a change, interpretation can continue; otherwise, the process terminates.

In summary, the source of knowledge is phenomena that appear as stimuli and what the input information may stand for according to our interpretation. The phenomenon indicated by the stimulus may not be uniquely determined. This non-determinacy explains the vast expressive power of signs, and it is also a primary source of confusion in sign determination.

1.2.3 Inherent Ambiguity

As a sign, the input quality can denote itself and also can stand for something else. Such an object, 'other' to the input (or stimulus), can be found through memory information triggered by the input. This information can refer to phenomena related to the input, such as train noise and a warning of danger. The important conclusion is that the concept of a sign is ambiguous in the above sense.[12] *Sign ambiguity* can be operational as long as the sign is subject to processing, that is, is in the 'making,' so its meaning is not yet finished (when a sign has already been interpreted, there is no doubt about its meaning, so no ambiguity either).[13] A specific feature of sign ambiguity is that the sign's meanings, in itself and as 'other,' are connected. For example, train noise (stimulus in itself) underlies a warning of danger by an arriving train ('other' meaning), which, in turn, assumes the existence of train noise.

The ambiguous nature of signs can be a source of misunderstanding, a welcome phenomenon in a joke, but this is not what we are after. The fundamental importance of this feature lies in its potential to improve the efficiency of information processing. Let us retake our example of train noise. Even if we cannot immediately recognize the input (i.e., in matching mode), additional information about the frequency components of the perceived sound, that is, information about the input itself, can make it possible to recognize the input sound as a kind of train noise, such as a warning of danger. In short, sign ambiguity can be the key to information organization and representation. A practical application of this feature of sign processing can be found in the music industry. The frequency components of a signal can be determined by using Fourier analysis. This technique is used in various equipment, including graphic equalizers. Audio devices facilitate dual processing by simultaneously producing sound and displaying frequency components, intensifying the musical experience.

1.2.4 Complex Phenomena

Linking the stimulus with memory information cannot always be arranged in a single operation without the danger of referential chaos, i.e., mixing up which memory information refers to which input quality (this confusion is related to the problem of determining which input subset represents agreement-sense and which only possibility-sense information, discussed earlier). Therefore, we assume that, in the case of complex stimuli, the collection of input qualities is processed in subsets. Only when those subsets are processed individually, as parts, and their data combined into a single relationship can the input be recognized as a whole. The need to combine information is also known from analysis mode processing. Since input

---

[12] That is the ambiguity between the sign as a quality and a reference to something else. Ambiguity due to memory information is not in our focus. Stimulus-level ambiguity, such as Rubin's vase, falls outside our theory (https://en.wikipedia.org/wiki/Rubin_vase). See also the footnote immediately below.

[13] Sign ambiguity can be resolved at the level of sensory perception and a higher level of cognitive processing. The first may explain why low-level sign ambiguity is not typical (we usually know what the input denotes). Zou et al. (2020) report evidence for disambiguation at the level of perception.

processing is intertwined with memory use, we have reason to assume that memory information is stored in a connected data network.

Network organization is advantageous, if only because it allows linking knowledge elements that are not directly connected via transitive closure. An analogous operation, in Aristotelian syllogistic (Aristotle's theory of deductive reasoning), is the reduction of a pair of propositions (premises) connected through a common term to a single proposition (conclusion). A necessary condition for a transitive closure is that information elements representing similar knowledge are structurally connected. Accordingly, agreement-sense response elements can be seen as nodes directly available for processing; possibility-sense response elements are linked to those nodes functioning as a hub.

As a result of network organization, relations between input subsets can be revealed by establishing a relation between memory elements. In matching mode processing, such a relation is available immediately. Analysis mode processing is more laborious. It requires selecting an element from the set of memory information that responds to the stimulus in the sense of possibility. In the worst case, interpretation must visit every response element individually. What is needed to find memory information efficiently in information processing?

1.3   Memory Organization

It is commonly believed that information is stored in memory by neurons. Individual neurons are connected through dendrites and axons, a neuron's input and output edges. The brain works massively parallel and can process multiple kinds of information simultaneously. Still, it can only focus on one thing at a time. In technical terms, human processing is single-threaded.[14] In line with this feature of the brain and the earlier consideration that similar knowledge elements must be structurally connected, we assume that complementary information is explored sequentially. As a result, knowledge elements encountered earlier during the search can be found more 'interesting' than those visited later. This way of processing implies the existence of an induced *ordering*, such as the ordering defined by the network structure of memory, which sets additional conditions on knowledge organization by the brain.

A mathematical concept that complies with the above assumptions is known as a graph. Formally, a graph can be defined by a set of vertices and a set of edges. A vertex is a knowledge element, and an edge connects a pair of knowledge elements. We assume that, in memory, as a graphical structure, vertices that respond to the stimulus in the sense of agreement exhibit high-intensity activation and function as a hub; those that react in the sense of possibility, a low-intensity activation which, in addition, decreases with the distance of an element from the hub to which it is

---

[14] https://theness.com/neurologicablog/index.php/humans-do-not-have-multi-core-processing

Fig. 1.2 Memory
activation. A black/white/
dotted-line bullet indicates a
high/low/zero-intensity
response; an edge denotes a
connection between
information elements

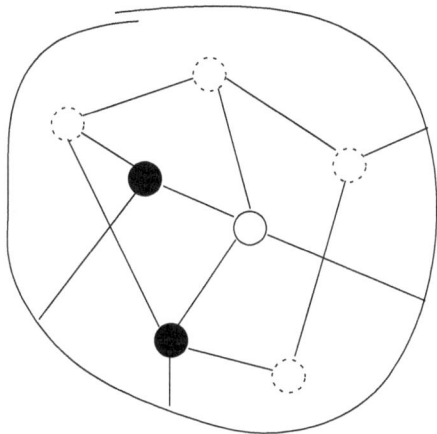

connected. See Fig. 1.2. It is like a nighttime image of a city. Buildings in spotlight
and shadow stand for high- and low-intensity memory responses, respectively. Dark
regions illustrate elements not responding to the stimulus. Buildings in the spotlight
can be found easily; those in shadow can be found by following a network of streets
starting from one of the illuminated buildings.

Can such an organization of memory arise? Since there is a permanent stream of
stimuli that offer themselves for interpretation by the brain and similar input
information can occur randomly over time, efficient memory organization could be
hopelessly complicated. And yet, there is no chaos in memory. The brain can
organize information optimally, enabling similar knowledge elements to be activated
by a single stimulus. It has been shown that unexpectedly, self-organizing memory
can be realized based on elementary statistics (Ritter and Kohonen 1989).

1.3.1   Memory Access

Besides storage size, memory access is an equally important characteristic of the
brain as a database. In this regard, two methods play a role: random and associative
access. The first allows access to any location in memory 'by address'; the second
provides access to the stored data 'according to content' without making any
assumptions about the location of information, which can be anywhere.[15] It is
generally accepted that information processing in the brain is based on associative
access. If so, what justifies the assumption of a network organization of memory?

---

[15] Associative access assumes memory response as an electric signal (cf. brain wave). Data
matching the external trigger can be determined by analyzing response wave characteristics, such
as frequency and intensity.

Let us start with the information processing conditions set by content-based memory access. Data responding in the sense of agreement can be treated as a single value; possibility-sense response information defines an induced ordering of values. The two representations allow for the two modes of information processing, matching, and analysis. The first, matching mode processing, is trivial. When processing in analysis mode, systematic search demands memory items already selected to be memorized. This condition may explain why memory information must be organized in a connected structure. But there is more to it than just making a to-do list.

The question is, what does the brain have to memorize and why? First, it must remember the external input to learn the types of external stimuli. Second, it must retain the response information triggered to understand its relative meaning, that is, the meaning of memory data from the perspective of the input processed. Conceptually, there are two ways to remember information. The first is storing the input as data; the second is establishing new links, thereby transforming the memory structure. The combination of the two methods allows for learning, including the introduction of abstract information through generalization. Since learning is assumed to be operational offline, the absence of the external trigger forces the brain to organize related data elements in a linked structure.[16]

While computer memory is only a storage medium, brain memory can change its structure in addition to storing data. Data that are similar or connected for other reasons are stored close together in memory. By considering connection strength as a measure of distance (the more their information is related, the shorter the distance between them in memory), we arrive at the model suggested earlier and used throughout this book. See also Fig. 1.2.

Associative-access-based dynamic and structure-based static memory representations can be isomorphic in terms of how a link between stimulus and memory data is determined.[17] Based on this relation and because, for our goals, structural properties of memory are only relevant, we will treat memory as a graphical network structure. We assume that memory matching the stimulus above a threshold, thus actively responding to the external trigger, is directly available for processing. Its information defines the focus of the interpreting system. Low-intensity memory response, passively reacting to the stimulus, thus in the sense of possibility, is available via links only. That information is not in the brain's focus and is complementary.

---

[16]Abstract concepts arise from instances through learning. Those instances must be close to each other in memory to facilitate generalization. This assumption is supported by lateralized language, which dominantly uses the brain's left hemisphere. Using both hemispheres would significantly burden the connection between the stored data (Ries et al. 2016; Ocklenburg et al. 2018).

[17]In associative-access-based representation, establishing a link requires searching in a frequency (value) space defined by responding memory data; in structure-based static memory representation, the same operation can be realized by searching in a graph.

In addition to visualizing data dependency, network organization is also helpful for another reason, as it enables the comparison of connected knowledge elements. How does this feature affect the use of memory response information?

Again, we may find an answer in sign ambiguity, the possibility of a twofold meaning of qualities functioning as a sign. The first is when memory information that responds to the stimulus is considered in itself, thus as a *value* or fact. An example is the input sound, interpreted as train noise. The second is when the stimulus is viewed from the perspective of something other than itself, thus as a *relative value*. An example is again the input sound, interpreted as noise from a fast-moving train this time. To this end, the input sound must be recognized as train noise and as a specific instance of this kind of noise, known as fast.

Does the brain facilitate the above two types of information? To answer this, we must delve into how knowledge appears in memory, which we will discuss in Chap. 2. Before that, we consider other practical questions of memory representation.

1.4 Abstract Phenomena

The examples examined so far relate to phenomena whose meaning can be easily grasped. What can we say about abstract phenomena and creations by human fantasy? Are they processed differently?

Abstract phenomena are phenomena with an abstract meaning. So, it must be a matter of interpretation. However, perceived qualities are always concrete, never abstract. We can only arrive at an abstract meaning by recognizing the input as it appears and then inductively generalizing its information. Abstraction is a hallmark of mathematics. Since thoughts are internal to the interpreting system, recognizing the perceived sound as a warning of danger and solving a mathematical problem could proceed similarly. Math problems also appear as quality; if the problem is not trivial, a solution can be sought using stored knowledge. Solving the problem may require induction, especially when proving theorems, but this[18] does not affect how an answer (cf. solution) is obtained. Moreover, the goal set in problem-solving is the same as in information processing: to test whether the stimulus means something essential and what an environmentally significant response would be

That mathematical problem-solving is indeed no exception is illustrated with an anecdote about Kurt Gödel, one of the most influential mathematicians of the twentieth century. According to the story, one of his students asked him about the best open problem in basic mathematics. Gödel replied: "How can we know if the problem is interesting." In terms of the theory of this book, the question is how to tell if a problem can be meaningful from some perspective. The bottom line is that while the complexity of our problems can be very different, finding a solution is always the

---

[18] The act of induction, thus not inductive reasoning.

same. A common element in all problem-solving is an emerging theorem (phenomenon), which is translated into a solution or proof (response), utilizing available rules (stored knowledge). That this holds for the best open problem in mathematics is nevertheless unexpected.

If mathematical problem solving is no different from information processing of any other input, then it is not inconceivable that mathematical problems occur to us like any stimulus. Hence, the assumption of a real mathematical world can be valid. How realistic this hypothesis can be is illustrated by an event from personal experience. During a scientific meeting in the 1990s, two fellow professors debated a proof of a theorem on vectors. They were in disagreement. At some point, one of them drew a picture on the blackboard and said in that questioning tone, "If you rotate this vector clockwise …" to which his colleague immediately replied, "Oh, now I get it." Although the theorem was abstract, the argument was concrete.

1.4.1   What Is Representation?

The fact that abstract mathematical and concrete diagrammatic knowledge can effortlessly be combined into one meaning indicates that their representations may not be different. This is not the first time we use the notion of representation, but what exactly does it mean?

As a verbal concept, representation is synonymous with interpretation, that is, determining the meaning of something or an action, for example, solving an abstract math problem. As a nominal concept, it can be anything that stands for something else and replaces it. For instance, the image of a vector can represent anything resembling a vector, such as a physical force, an increase in price, or other value. As a nominal notion, a representation can also stand for itself, but usually, that is not the intended meaning. An example is a vector, seen only as a form. Representation is, therefore, also a synonym for sign. In short, the notions, sign, representation, and interpretation denote different views on meaningful expression.

1.5   Information Merging

Information processing must deal with heterogeneous qualities. Partly, this is compelled by the brain, as an interpreting system, which is fed simultaneously with information from multiple senses and processes that data based on prior knowledge, and partly by the need for a meaningful response, where the input must be viewed from different perspectives, and thus analyzed by using different types of memory information. Yet the brain, as an interpreting system, can merge those disparate data easily and transform the arising collection into a single response.

For example, the driver can interpret the appearing sound as a train sound (from a frequency perspective), as a warning (from a semantic point of view), and as input entering his ear from a specific direction (from a topological stance). If the signal points at the level crossing, the driver may take the sound as a warning to stop the car. His response will be different if it points to the car radio he is listening to. In the same vein, only if the driver can process information by the traffic board, "WAIT," as a well-formed phrase (syntactically) and also as a warning of danger (semantically) that he will stop the car in response.

In sum, stimuli can be viewed from various perspectives. A perspective can be understood as a type of information or quality, thus a property, and a collection of qualities corresponding to the perspective as a *knowledge domain*. Merging abstract mathematical and diagrammatic knowledge assumes the existence of shared qualities that characterize their information, that is, a perspective from which their meaning can be compared. What perspectives are there? How do they emerge? To answer these questions, we need to know more about the putative way of knowledge representation used by the brain, which is left to the next chapter. Before that, we will look closely at a related issue: the conditions for merging information. The possibility of common qualities is undoubtedly one of them. The question is, what does it take to realize it?

Stimuli can elicit a response in different knowledge domains. Because of the need for a fast response, information triggered by the input must be efficiently put together into a comprehensive representation. If the driver cannot combine information about his car's location and the warning of danger, he will keep driving or stop the car unnecessarily. Which of these responses will eventually be obtained can depend on other conditions, such as the driver's experience with level crossings or his fear of an accident. The brain can efficiently combine information, but how it does this has yet to be revealed. When we want to implement the same functionality by the computer, we realize the complexity of the underlying problem. Here is an illustration.[19]

Assume you ask your search engine how to dry your clothes, and the answer is "Hang the clothes in the garden." Your next question concerns the weather conditions, and the system tells you that "it will be raining soon." At this point, a human agent can quickly draw an obvious conclusion. However, it is unlikely that your search engine could combine its responses into a single piece of advice, "it will rain, so you should not hang the clothes outside." It is not that the above problem and, from a general point of view, information merging cannot be solved. In a computer setting, this requires a series of programs that translate information between different domains, for example, information about the weather, into conditions for drying clothes. Having the necessary programs is by no means trivial. For n knowledge domains, the number of translators is in the order of $n * n$. A solution can be feasible for low values of n. However, for ever larger values of n, the problem can become unmanageable.

---

[19]This illustration is based on an example introduced by Peirce.

Fig. 1.3 Sample portability problem (S/T/CL = source/target/common language)

1.5.1 Universal Common Language

The idea that information merging can be drastically simplified by using a uniform representation for data first popped up in the 1950s in compiler construction (Strong et al. 1958). Even today, this concept is interesting because it shows the limitations of a formal knowledge specification. Compilers are programs that translate texts in a source language to code in a target language. Only if the target language is implemented on a computer can the output code be executed immediately on the same computer. Otherwise, the code must be ported to a computer that can run programs in the target language.

How computer science tried to solve the problem of multilingual translation and eventually failed is illuminating. Hundreds of programming languages were designed in the late 1950s and the next two decades and many were implemented. Efficient programming, including a natural formulation of algorithms, was central to computer science research. At the same time, numerous new hardware architectures, in particular processors, were put on the market. They had a translator for their low-level assembly language and compilers for a few high-level programming languages. Compelled by the high program development costs, computer science was burdened with translating between many architectures and languages, heralding the infamous problem of program portability.

The idea of a mediating language has been launched to address the portability problem. Using a single common language can reduce the number of translators from $n * m$, which can be high to the lower value of $n + m$ (where n and m denote the number of source and target languages, respectively). As a result, instead of developing translators between all languages, only compilers between each language and the intermediary common language are required. See Fig. 1.3.

At the outset, everyone found the universal common language an attractive idea. Yet, it has never been put into practice. What was the problem? Using a common language can reduce the number of translators, not the number of languages and hardware architectures. The compilers had to be written in the universal language for portability reasons. This language had to be of a high level to facilitate abstract programming, which is necessary for a transparent formulation of complex algorithms; to access the hardware architecture, a requirement imposed by the need to generate efficient code, the language also had to be low-level.[20] The conflicting requirements resulted in the definition of a programming language, which was challenging to use, let alone implement.

Program portability proved to be a severe issue at that time. Driven by the software industry, the number of hardware architectures and programming languages declined significantly in the 1980s and beyond, reducing the impact of portability on software development to acceptable levels. Notably, not the most elegant, but the most practical languages survived, just like in natural evolution, but this is another story.

Unlike programming languages that are formally defined, natural language offers better opportunities for multilingual translation. By understanding the meaning of language utterances, human translation can be highly flexible, using metaphors and language variations. The downside of human translation is that it is not automatic. The ease of translation between natural languages makes the hypothesis hopeful that human processing could rely on a uniform representation of information.

1.5.2 *Uniform Representation*

Translation between languages and linking knowledge elements from different domains needs information about the compatible source and target concepts. Such a lexicon is the key to meaning-preserving translation and merging data. Assume the task is to translate 'Evening Star' into German. Intuition suggests that this should be 'Morgenstern' if the textual context refers to the time of day and 'Venus' if it relates to planets. To solve this problem, we need information about time, planets, and other related properties. Furthermore, we must combine all this information to select context-dependently a correct translation.

What is required for efficient merging? Corresponding information elements must have an identical type to eliminate the need for translations. However, more is needed. The relations of those elements in their respective domains must be respected in the same way. Again, this is when a transitive closure operation comes in handy. 'Evening star' can be linked to information about the time of day, that Venus appears before sunset and stays for some time after sunrise; 'Morgenstern' to data that daybreak precedes midday, and possibly also to information about the German poet Christian Morgenstern, and the list is far from complete. Competent translation might prefer 'Morgenstern,' not 'Venus,' to facilitate the later sarcasm, "also musst du früh aufstehen" (so you should wake up early). Making the correct choice presupposes knowledge that "Morgen" and "früh" designate similar moments, i.e., comparable time measures.

---

[20] A high-level concept is a procedure as an abstract control structure. A low-level one is a hardware mechanism for passing procedure parameters.

Fig. 1.4 An illustration of nesting

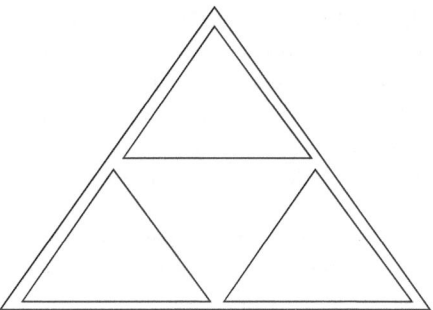

We can only combine concepts from different domains if they are compatible from some perspective. Verifying this condition can be demanding unless knowledge is represented uniformly. In that case, translation can be replaced by a structural combination, which is easier to handle. This way of information merging also underlies syntactic coordination in natural language. An example is an utterance, "Venus is a planet, and it shines at night," in which "Venus" and "it" and "a planet" and "shines at night" are subject to coordination, i.e., syntactic merging, based on the underlying language structure. Although the relation between those phrases has yet to be verified, the example illustrates that the first step in syntactic processing can be accomplished relatively simply.

Earlier, we noted that phenomena that cannot be immediately processed can be interpreted recursively using nested processes. The idea of nesting is geometrically illustrated in Fig. 1.4. Three instances of a (small) triangle recur in an embedding (large) triangle. Recursive processing always uses a single type of representation (see a triangle). In natural language syntax, a structure that can occur recursively is a closed clause. An example is the sentence, "(The planet known as Venus) shines at night." The closed clause, in brackets, functions as a nominal (subject) in the clause of the entire sentence.

Can we assume that knowledge is represented uniformly by the brain? There is no direct evidence for this hypothesis. However, indirect evidence can be found in neurophysiological research on language processing. Meaningful processing of language utterances requires that they are syntactically and semantically well-formed. According to experimental research, syntactic and semantic analysis simultaneously occur in the brain (Hagoort et al. 2004). Evidence shows that a temporal difference between information processing in those domains is significantly smaller than the time required by the two kinds of analyses processed independently. This difference in time implies that the hypothesis of sequential processing—syntactic followed by semantic– and thus information merging through translation may not hold (in these two domains, at least).

1.6 A Single Type of Process

The possibility of uniform representation of information reinforces the existence of a single type of process for interpretation. If that exists, the conjecture of a common element shared by all interpretation processes must be correct trivially. Motivated by the fallacy of the idea of a universal language, we argue that this common element must be more elementary than a language. It must be a type of process. The process must be domain-independent to apply to all kinds of data. How can such a type of process be defined? How can the condition of domain independence be met?

Let us begin with the first question. Here, the key is the concept of abstraction. An example in geometry is the introduction of the notion of an angle. Instances of an angle have the same meaning in all polygons, regardless of their shape and size. Abstraction also applies to interpretation processes. One way to get there is to realize that interpretation events use data. To define interpretation as a type of process, we must abstract events on individual data domains into events on data types, which brings us to the second question above.

Following the idea of interpretation as tiling, we suggest that domain independence can be attained through generalizing quality co-occurrences in types of relations. To this end, more than a uniform representation is needed, as it is limited to the structure alone. Classification of quality relations asks for a theory of interpretation. In Chap. 4, we will argue that Peircean theory provides such a framework. For now, we focus on the relational meaning of qualities concerning the problem of domain independence.

One way to achieve domain independence is to abstract qualities into combinatory types. The question is, what types are there? The existence of combinatory properties is known from language. An example, in syntax, is the phenomenon of a noun-adjective modification, generalized into a relationship between an active combinatory need of the adjective (modifier) and a passive combinatory need of the noun (argument). Based on this abstraction, instances of adjective-noun modifications can be combined into a single structure if their constituents are type-wise compatible.

Introducing abstract combinatory needs brings efficient, in technical terms, low-complexity information processing within reach. The hypothesis of efficient interpretation is consistent with Chomsky's suggestion that the language faculty of the brain contains innate capacities for each language and must, therefore, be simple. Although domain-specific knowledge (cf. mental lexicon) can be voluminous, thanks to the single type of process and a uniform representation of information (based on abstract combinatory properties), meaningful information processing can be simple and efficient.

Returning to the first question, namely, how to define such a process, we must admit that it is too complicated to answer immediately. To get there, we need to take a few hurdles first. Because some of them are about technical issues and not directly relevant to the subject of this chapter, we defer their discussion until later. For now, we assume that such a process can be defined. Building on this, we proceed with

analyzing the conceivable consequences of the hypotheses of a uniform representation and one type of process, which is more intriguing.

1.6.1   Terminology

Before we go any further, we must clarify the used terminology, particularly the
concept of process and type. The meaning of a *process* is more sophisticated than
one might expect. It is commonly known that a process is a series of events. On the
other hand, it is less well-known that such a series can only be considered a process if
the individual events are coherent from a particular perspective. Suppose you are
traveling to Amsterdam. You buy a ticket, get on the train, sit, and watch the
picturesque landscape as you travel. Assume that you change your mind midway
and step out of the train. Do the above events define your traveling to Amsterdam as
a process?

The answer is negative. A process is a sequence of events governed by a goal. If
you leave the train midway, you have not reached Amsterdam, your goal. Therefore,
the process of traveling to Amsterdam does not exist either. This condition also
applies to information processing. If the processing does not reach its goal, there is
no response. Thus, it is similar to not noticing the input at all.

Now, let us take the second concept, the concept of *type*. We can generalize
individual events into a kind of event. An example is the event of taking a seat on the
train. By disregarding the type of seat (e.g., window or aisle), this event can be
generalized into a kind of event consisting of sitting down, if the seat is in the aisle,
and making a few steps and then sitting down, if it is by the window. Also, we can
abstract a process into a type. Traveling to Amsterdam by train can be generalized as
traveling to another location using some transportation. In programming languages,
a process can be implemented by a procedure, a type of process by a generic function
that applies to different data types.

1.7   Interpretation Versus Computation

It is commonly assumed that interpretation is qualitatively more complex than
computation. Results of an experiment by Vilayanur Ramachandran and William
Hirstein (1997) indicate that this hypothesis can be correct. This experiment displays
an image that subjects first saw only as a collection of spots.[21] By looking at the
image for some time, they recognized the shape of a Dalmatian dog. It is fascinating
that the subjects first saw the input image only as a random pattern (that is, before

---

[21] https://franzcalvo.wordpress.com/2014/07/08/richard-gregorys-dalmatian-image/

recognizing its meaning). Once they noticed the shape of a Dalmatian, they could never again see the input as just a collection of spots.

Interpreting the image for the first time can be laborious. Still, once we are familiar with its meaning, we can immediately recognize all later occurrences of the same input. This ability is consistent with the assumption of two modes of information processing: analysis and matching. The importance of this distinction is secondary, however. The crucial conclusion of this experiment is that interpreting the input as random patterns and a meaningful concept must be qualitatively different. Interpretation is more than just pattern matching. It is a transformational process.[22]

In the digital era, the question is *not* in what sense human interpretation is more developed than simple computation. Instead, the question is what it takes to represent the world meaningfully using the computer. In other words, what sort of calculations by the computer can be easily experienced as meaningful and so can augment human intelligence. In everyday life, we are forced to use computers for efficiency reasons. According to a hypothesis of this book, the efficiency of computer use and confidence in its results can be increased if the computer calculations respect the properties of meaningful processing.

1.7.1 Ubiquitous Interpretation

The conclusion above brings us to the question of who/what is capable of interpretation. Could it only be the human brain? Traditionally, interpretation is understood as an act of explanation of something or some property by a human agent. In a broader sense, we use this concept as the ability to respond to a stimulus. Accordingly, animals, plants, and all living creatures must be able to interpret.[23] Sunflowers react to sun rays by turning in their direction (the French word 'tournesol' clearly conveys this meaning). Also, physical phenomena can be said to possess a capacity for interpretation. An illustration of this is the ability of mass to respond to the existence of another object with an attraction proportional to the mass of the two of them.

Both of the above examples assume the use of a rule. This rule is heliotropism, in the case of the sunflower, and that of gravitational force, in the case of mass. Human interpretation typically uses rules arising via abstraction, i.e., generalization. It is not unthinkable that laws of nature also have a similar character. An illustration of this is the conductivity property of metals such as iron. Iron atoms do not conduct electricity. Electrons loaned by iron atoms to the crystal structure of metallic iron are what conduct electricity. As long as the number of iron atoms forming a crystal is

---

[22] And, thus, related to a qualitative change.

[23] Jellyfish have demonstrated that the existence of a centralized nervous system is not necessary to learn by association. See https://www.pnas.org/doi/10.1073/pnas.2220685120

around two, conductivity, as we generally know it, cannot be observed. With an increasing number of atoms, this property arises almost instantaneously, as if by a qualitative change. Accordingly, this and all larger collections of iron atoms can generate electrical current under voltage, thus exhibiting conductivity as a rule-like property.

1.8   Towards a Theory

After this introduction, let us return to our central problem, interpretation, and how we think a solution for a computer that can think human-like can be found. Interpretation is related to the fundamental concept of the digital era, namely information. Many problems have a bearing on that notion, but the focus of this book is limited to information as a process. Central in our inquiry is the assumption of a single type of process and a uniform representation of knowledge. How can these hypotheses be proven?

A standard view in the twentieth century and beyond is that knowledge can be defined as a set of facts. Here, we take a different standpoint by suggesting that knowledge and meaning arise through a transformational process. A framework supporting this view can be found in the work of the American polymath Charles Sanders Peirce (1839–1914). Based on his theory, we introduce a model of meaningful information processing that suits a computational implementation.

1.8.1   Peirce

Peirce was an extraordinary scientist, unparalleled in history. Today, he is appreciated for his contributions to logic, mathematics, philosophy, and the foundation of pragmatism, among others. He is also known for his theory of the formalization of reasoning, which he developed independently of the German logician Gottlob Frege. Despite his fundamental results, Peirce did not receive the recognition he deserved, at least not in his home country. He was the target of vicious forces, which made his career impossible. How Peirce dealt with his problems was not successful either. Peirce died in poverty, for which he was, at least partly, himself responsible.

Why are we interested in his life? For the same reason as, for example, why we may find it necessary to visit historical places. Factual information alone is unsatisfactory. We need to see the site where a particular event took place. A similar confrontation of fact and a subjective impression is recognized by philosophy in the idea of a sublime historical experience (Ankersmit 2005). From the stance taken by this book, such an experience enables combining factual information with information from subjective observation. As a result, later similar stimuli can trigger a richer memory response and, thus, a more intriguing input meaning. Familiarity with Peirce's life can affect our thinking in the same vein.

Chapter 2
Knowledge Is Formed by Cognition

Abstract Imagine the brain as a complex system that continuously receives and processes information. This process has two main sources: the stimulus (information you are perceiving now) and memory (data you learned from previous experiences). How is the stimulus represented as input information? How is the result of information processing stored in memory? What can a network organization of memory data explain? We will argue that, in the input interaction, the stimulus affects the brain/mind and changes its state. Accordingly, sensory information is a relation of state and effect data. Conceptually, the organization of network memory allows for a similar interpretation of stored data. Memory elements that are connected can be subject to comparison. Data shared by, and that are different, can be interpreted as state and effect relation. The identical interpretation of input and memory information allows for a uniform data representation.

2.1 How Is Knowledge Stored?

Limitations in memory can be attributed to two factors: the conditions of interpretation (how memory storage is organized) and the properties of knowledge (how stored knowledge is represented as data). After briefly discussing the first in the Introduction, we analyze the second in this chapter.

Input processing has two sources: stimulus and memory. The first refers to the phenomenon perceived now; the second relates to phenomena observed in the past. The meaning of input qualities as a value or property is obvious. Memory information, which can be abstract, is more difficult to explain. An example of a stimulus is the visual signal of a chair, informing about color, shape, size, etc., as qualities. An illustration of memory information is the concept of a chair, which refers to any object that qualifies for 'something to sit on,' such as an armchair, table, couch, windowsill, and even a toilet seat. We can have memories of many such objects observed in the past, and the magnitude of their data can be enormous.

Although the scope of a memory concept can be broad, the set of objects it denotes can, in principle, be enumerated. This operation may require extreme concentration, but this is not something that a human agent cannot achieve. The

© The Author(s), under exclusive license to Springer Nature Switzerland AG 2025   23
J. J. Sarbo, *Inevitable Knowledge*, SpringerBriefs in Computer Science,
https://doi.org/10.1007/978-3-031-73461-8_2

funny thing is that listing the contents of the memory is a trivial task for the computer. If our original problem is not a consequence of the large extent of memory concepts, then it must be due to the structural properties of stored information. This thought brings us to the question: how does the brain represent knowledge?

2.2 Typical Representation

The brain is primarily a continuous processing system (Tee and Taylor 2020). Folded into real-time information processing, several levels of discrete processing exist. The need for discretization can be justified from an informational stance. Since stimuli can occur to us arbitrarily often, but interpretation can take time, input activation must be kept invariant during information processing to get a meaningful response. A distinguishing feature of discrete processing is the sampling of the continuous input in a collection of qualities. We use the terms collection and set interchangeably, but this is far from obvious. In mathematics, a set consists of elements that are unique and accessible via selection;[1] a collection may not meet these conditions. Since data on information processing at the level of neurons is not available, we cannot know whether memory information is stored uniquely and can be retrieved individually, which justifies the synonymous use of the two terms.

Memory organization is a matter of storage size. With unlimited memory, all information can be stored; with finite memory, efficient storage usage is unavoidable. As for the human brain, which is known for its staggering capacity of 10^{11} (hundred billion) neurons, storage size is not an issue.[2] If each neuron could only store a single piece of information, the brain had only a few gigabytes of storage space. However, neurons are connected to other neurons and realize a network of more than 10^{14} (hundred trillion) synaptic connections. A single neuron can help store many memories at a time through combinations with other neurons, increasing the brain's capacity to around 2.5 petabytes (million gigabytes). This memory size would be enough to store three million hours (roughly 300 years) of digital video recording.[3]

While memory capacity can be satisfactory, access to information activated by the stimulus may be limited. On the one hand, access to agreement-sense memory response information should allow for a prompt reaction as stimuli can occur in quick succession; on the other, if immediate input recognition is not possible, possibility-sense response information (including data that are in some way related to the stimulus)[4] must be available for some time to allow selection from the vast collection of complementary data. Why is a selection necessary? Interpretation is

---

[1] https://en.wikipedia.org/wiki/Axiom_of_choice

[2] https://www.scientificamerican.com/article/what-is-the-memory-capacity

[3] https://pubmed.ncbi.nlm.nih.gov/15925809

[4] E.g., data enabling hypothetical input processing.

deterministic, and there is no place for rehearsal. However, most of the time, the stimulus does not precisely match the stored information. Thus, agreement-sense matching is only possible for a subset of the input qualities. Accordingly, in analysis mode, the interpretation process has to visit the elements of the memory response matching the input in the sense of possibility, that is, below a threshold, to recognize the stimulus in its entirety.

The assumption of threshold values implies an organization of memory information based on types. Data that respond to an input quality with an intensity above a threshold are considered equivalent and thus of the same kind. The hypothesis of a *typical* representation of knowledge is consistent with the theory by psychologist James J. Gibson. Gibson claimed that stimuli are more than arbitrary qualities. In his view, meaning is rooted in 'affordances,' i.e., what the environment affords the observer (Gibson 1979).

Affordances stand for opportunities for action provided by a particular object or environment. They appear as qualities that are potentially useful for our goals. Their values can vary within certain limits, which explains the relationship with types. Donald Norman called them action possibilities,[5] that is, how an object may be interacted with. The perception of a hammer can illustrate affordances. This object can be grasped in many hand poses. Still, there is only a limited set of effective contact points (the values of which can vary between limits) and their associated optimal grip for using the hammer efficiently.

Although knowledge must be organized into types for efficiency, Gibson's theory suggests that those types are at least partly forced upon us. A concept related to affordances is known in cognitive theory as selective attention. Knowledge required for selective processing can come from learning, but it can also be innate.

2.2.1 *Value and Measure*

How is knowledge represented? We must return to the properties of memory organization to answer this question and the pending question about a dual interpretation of the stimulus as a factual and a relative value.

The need to remember past stimuli stems directly from the goal of input recognition, which says that input and memory information form the basis for determining the response of the interpreting system. This idyllic picture is disturbed by temporal conditions. One is the possibility of an occurring more critical stimulus, and another is the danger that the intensity of the current trigger, and thus the intensity of the memory response, falls below a critical value. Both conditions reinforce the importance of efficient information processing.

Because the collection of response information can be extensive, the brain may not have sufficient time to look at each information element individually. We can ask

---

[5] https://jnd.org/the-design-of-everyday-things-revised-and-expanded-edition

Fig. 2.1 A sample linearly ordered (left) and an unordered set (right)

ourselves why would it want to do that. Why is it not enough to take one response, for example, the one with the highest intensity? The reason is that the input can elicit multiple responses of similar intensity that are equally suitable for information processing.[6] Their data needs to be treated as a whole for a quick response. How can data collections be dealt with as a single value?

There are two obvious solutions to this problem. The first is when a data collection is represented as an *average value*. Here, the term 'average' denotes a general meaning, such as the concept of an 'average' man, and not a statistical average. An analogous concept known from cognitive theory is a prototypical value. The second solution is more advanced. While an average value can represent a data collection as some 'thing' or object, it may not inform about the property the elements share. The existence of such a property requires the collection of elements to be compared with each other, which assumes that they have shared qualities.

Enumerating all elements of a set is facilitated by the network memory structure. Deriving a property can be efficient when the set is arranged in an order. In that case, the difference and, generally, the order relation between the elements can be used as a property. However, a property alone is insufficient for a collection of data seen as a relative value. It is also required that the stimulus be recognized as an instance of this property. The problem of which response element the stimulus is closest to can also be solved using comparisons.

An illustration of the principle of how this all works is as follows. Consider a memory that is a store of natural numbers. Assume the stimulus is the value '4' and the memory response is the ordered set {2, 4, 6} and all set elements respond in an agreement sense, with element '4' showing the highest intensity. See Fig. 2.1. Consecutive elements of this ordering differ by a value of 2. The property that the elements share is an 'even number.' The input value '4' is an element of the set, thus satisfying this property.[7]

To determine membership and infer a shared property proceeds differently when the memory response is an unordered set, such as {6, 2, 4}. In this case, we cannot assume that the stimulus is not in the set if we encounter a larger value. Also, finding a property may take longer, and it may also fail. And it is precisely time that is limited available for input processing. Thus, the question of which element of the collection is closest to the stimulus can only be answered efficiently if the elements are ordered somehow. Also, a property can be derived more easily and added as a quality to each element through learning. If the set is not ordered, determining

---

[6]These are the collection of agreement-sense memory responses in matching mode and the subsets of the possibility-sense responses of a similar measure of intensity (the set of measures defines an induced ordering).

[7]Knowledge of this property and its symbolic name is optional to achieve this result.

Fig. 2.2 Velocity as a value
and measure

membership and inferring a property requires exhausting searching, which may not
be feasible for large data sets. Data representation based on frequency and intensity,
assumably used by the brain, trivially allows for comparison and, thus, also the
representation illustrated above.

In short, the induced ordering allows the response set to be characterized as a type
and to position the input as a value of this type and also as a value relative to the
response data as a set. We call such an ordering a *domain* and such a relative value a
measure. A measure can be considered a value and, alternatively, an initial subset or
range of a domain, for example, the value '4' and the subset {2, 4}, respectively.

All qualities, not just numbers, can be represented as a domain and a measure of a
domain. An example is the input sound, recognized as the noise of a fast-moving
train. This means that the memory response is interpreted as a domain ('train noise')
and the stimulus as an element of this domain ('fast-moving'). This element can be
seen as a value and a subset of values less than or equal to the speed indicated by the
element's value. An analogous twofold interpretation is witnessed in natural language.[8] In language, values and measures can be denoted by identical and different
symbols. An example of the former is "move," in the utterances "a good move" and
"move a chair." In the first, "move" in itself is characterized by the property "good";
in the second, it is referred to from the perspective of "chair" as a measure of "move"
type actions (cf. chair-moving). An illustration of the use of different symbols is
"80 km/h" as a speed (value) and "fast" as a velocity (measure), assuming this speed
and velocity refer to the same element of their shared domain. Besides their value
meaning, they designate a range between 0 and 80 km/h. See Fig. 2.2.

---

[8]Language symbols appear as qualities and thus can be processed like any other phenomenon.
Because utterances can consist of a series of symbols, language interpretation can require a series of
processes. We return to this point in Chap. 5.

(2)   1 1|
(4)   1 1|1 1
shared   different

Fig. 2.3 Shared and different qualities of {1,1} and {1,1,1,1} in unary code

2.2.2   *Perspectives of Representation*

The concepts, average value, and measure of a domain are clear. A pending question is how their use affects the representation of stored information. To simplify the underlying problem, we revisit our number-memory example in the previous section.

The brain can perceive the stimulus as a number (we will discuss later what this enables). For the goal of illustration, we assume that number representation is based on unary code.[9] A specific property of this code is the use of a single symbol. For simplicity, we designate this symbol and the corresponding quality identically with '1.' For example, the natural numbers 1, 2, and 3 (in unary code, 1, 11, and 111) are stored by the collections {1}, {1,1}, {1,1,1}.

Having said this, let us look at the response set {2, 4, 6}, triggered by the stimulus '4'. As an average value, this set includes 2, 4, and 6 as values. In a sense, the stimulus involves the meaning of these numbers as prototypical values, as an 'average value.' Relative value interpretation offers more possibilities. The comparison of a pair of qualities, which are themselves a collection, can yield two kinds of results: qualities shared and those different. This ability for representation is when unary code comes in handy. By comparing successive elements of the set {2, 4, 6}, in unary code {{11}, {1111}, {111111}}, we can conclude that shared qualities by 2 and 4, and 4 and 6, are {1,1} and {1,1,1,1}, respectively; those different are {1,1}. See also Fig. 2.3. If we know {1,1} as an 'even' number, we have a (symbolic) property that characterizes all elements of the memory response. Combining the two meanings of the input, as an average value and a measure of a domain, we get a property of the stimulus: '4 is a number like {2, 4, 6} and possesses the property of even', or briefly '4 is an even number.'

To conclude that a particular property applies to every element requires reasoning. For sets with many elements, more than deductive reasoning is necessary; we need induction. Having too many elements does not mean the set is infinite. It means that it contains more elements than information processing can handle within the time available. The other way around, the existence of a property and successful testing of one or more items for this property could be the ground for a naive meaning of infinity and the ability for inductive generalization.

Hypothetical reasoning is needed to determine what other properties can be assigned to the response besides the property 'even number.' A nice feature of our number example is that it illustrates this thinking, too. Hypothetical reasoning can be

---

[9]Data representation based on intensity values is analogous to the unary code.

processed analogously to complex phenomena, that is, via recursion. Recursive processing treats the response by a nested phenomenon as a quality of a more extensive (embedding) phenomenon. Hypothetical reasoning, as a process, uses nesting in a different sense, however. Unlike regular processing, in this case, the interpretation process does not stop when the stimulus has been recognized and the first or immediate meaning of the input has been found. How is it possible? Why is it happening? The first question is simple. As long as the activity of the stimulus allows it, the interpretation process can continue. The second question is more complicated. Before discussing it, let us return to the conditions of nested processing.

Remember that the arising response, as a quality, may activate memory. Let us assume that the thought, '4 is an even number,' triggers a (new) memory response, e.g., the decimal value '10,' which is an even number too, thus similar to the current input. Based on the existing response set and this new value, the interpreting system can discover a novel property of the memory response. Accordingly, starting with the second element (the value 4), each subsequent element is the sum of the previous two values. Indeed, $2 + 4$ gives 6, and $6 + 4$ is 10. The new property also applies to the original response set ($\{2, 4, 6\}$). How the new information ('10') affects the interpretation of the sample input ('4') is beyond the scope of our example.

Because memory knowledge can be vast, the brain can develop numerous explanations for the external stimulus recursively. The essential conclusion of the above example is that the emerging properties can explain the meaning of the answer set and, thus, of the input from different perspectives. In short, the concepts of perspective and property can be synonymous.

Coming back to the second question (why hypothetical reasoning happens to us), we note that memory activated by '4 is an even number,' as a quality, can be complementary information already available, but also new information not used by the process so far (data responding to the stimulus in the sense of agreement has already been processed and therefore are not our concern now). Information by the first cannot contribute to new meaning. This condition does not apply to the emerging new information by the second, which initially did not respond to input but allows exploring new yet unknown reasons for the stimulus. If successful, this 'discovery process' may reveal explanations that have little to do with the intended meaning of the stimulus. The interpretation process can wander this way but may become sharp again when the stimulus reappears in our perception, for example, thanks to an involuntary movement of the eyes.

In contrast to processing complex stimuli, where nested phenomena are represented as a quality in an embedding phenomenon, hypothetical reasoning considers the meaning of the input phenomenon as a quality in the plethora of stored knowledge (cf. nesting phenomenon). This understanding of nested processing leads to an unexpected conclusion. According to this, besides responding to the external stimulus, the core activity of human processing is to reveal possible explanations of the input and thus discover the world as a set of phenomena. Observed phenomena can be learned as objects and properties by representing their data as an average value and a domain measure.

Investigating alternative explanations is no more than a possibility. Why does the brain learn new properties this way? Discovering possibilities is a hallmark of evolution. In Chap. 5, we will argue that interpretation may characterize all phenomena. If this assumption is correct, Darwinian evolution could be seen as a result of hypothetical reasoning by nature.

Meanings from different perspectives can also be derived by analyzing arbitrary subsets of the set of memory response information. To our knowledge, there is no evidence for this way of processing by the brain. Not so for the computer, for which access to subsets of stored data is simple. Would this option be attractive? Not in our view. By considering arbitrary subsets, the computer can 'discover' properties that may not refer to a 'real' phenomenon. Such formal properties may not inform about the world as it is and thus may be meaningless.

By now, we have all the ingredients for defining our process model of interpretation. Before that, we first tie up loose ends by reviewing open questions from the Introduction.

2.3 Reliable Calculations

One of the questions was whether we could trust the computer's response without any reservations. The answer is not unequivocally affirmative. For example, we need to understand how an algorithm works, especially when there are too many dependencies to account for. To increase transparency, AI uses statistics-based algorithms, such as Bayesian inference. With this type of reasoning, we can gain insight into the effect of an individual event (cf. evidence) on the probability of a hypothesis. Here is an example. Flu is often accompanied by fever. Based on statistical analysis data from flu cases, the chances of fever can be calculated. The novelty of Bayesian reasoning lies in the possibility of thinking in the opposite direction, from evidence for a fever to the risk of flu. This type of analysis can increase the quality of programs used in strategic games, expert and planning systems, and all sorts of measurement-based research. However, this does not mean human probability estimation and Bayesian statistics are identical. Experimental evidence shows that their estimates can differ widely.

Returning to our question, could our doubts stem from the feeling that statistical data analysis alone cannot provide valuable responses? Or, we trust computational solutions but do not think that current AI technology can exploit the computer's full potential? First, it is unlikely that statistical reasoning alone can achieve the complexity of meaningful processing, not because of an abundance but rather because of a lack of sufficient data. The amount of information processed by the brain's 10^9 neurons and 10^{15} synaptic relations is orders of magnitude greater than the amount of data available from measurements of neural activity. Measuring activity by voxels limits the granularity of current MRI, fMRI, and EEG techniques. These are three-dimensional spatial compartments containing thousands of neurons. In addition to this obstacle, brain signals suffer from noise, which can make neural measurement unreliable. Finally, the conceptual distance between neuron-level activity and mean-ingful interpretation is too great. Whether data from the former alone can be sufficient to explain the concepts used by the latter has yet to be determined.

A lack of confidence in the computer's response may also have another reason unrelated to statistical reasoning. A famous conflict between theories of planetary motion in the past sheds light on its essence.

2.3.1   A Historical Lesson

At the end of the seventeenth century, there were two competing models of the planetary system, one proposed by the Danish astronomer Tycho Brahe, a follower of the classical Ptolemaic view, and another by the German scientist Kepler, an advocate of Copernican theory. Brahe's model was Earth-centered; the Sun revolves around the Earth, and all planets revolve around the Sun. Kepler argued that the center of the planetary system must be the Sun. All planets, including the Earth, revolve around the Sun. The two theories also differ in the geometry of orbits. In Brahe's view, planetary orbits have the shape of a cycloid (see the curve traced by a point on the rim of a wheel as it rolls along a straight line).[10] According to Johannes Kepler (1571–1630), the orbits of the planets, including the Earth, are ellipses.

The question is why there was a preference for Kepler's heliocentric view. It should be noted that Kepler presented his theory only as a mathematical model. Kepler's explanation of why his theory is superior to Brahe's and Ptolemy's model was based on his assumption that mass always revolves around masses (McGowan 2009). Epicycles introduced by Ptolemy and Brahe, which are cycles around an empty space, do not fit this view. Today, such cycles cannot be thought of as orbits in a gravitational space. Kepler was right, of course, but it took half a century for Isaac Newton (1643–1727) to discover that planetary motion results from gravity as the attraction between bodies. Newton validated Kepler's empirical law through func-tional calculus, which he developed independently of Leibniz. Returning to Kepler, his theory has a significant advantage, namely the potential to model phenomena related to planetary motion but fall outside the realm of geometry. Most importantly, Kepler's elliptic orbits facilitate a model of revolution time, which is consistent with experience. A similar interpretation of Brahe's cycloids is far too complex and unnatural.

Let us note that Brahe's and Kepler's models both allow for a definition of revolution time. However, if we are concerned with this concept as an observable quality, not a formal notion, then the difference between the two models should be clear. Indeed, in addition to understanding an orbit as a geometrical concept, we may assign the points of its curve the meaning of successive moments in time. Regarding

---

[10] According to Brahe, if the planet's motion along the epicycle is faster than the epicycle's motion around the Earth, then the planet can appear to go backward for parts of each orbit (Kutner 2003).

the interpretation process, the two kinds of information, geometric and temporal, jointly contribute to the meaning of planetary motion. Since our responses should be to the best of our knowledge, we would better use a theory that allows for a natural combination of the two types of information. In Brahe's model, a planet seen as a point proceeding along a cycloid appears to move backward in parts of its orbit, which is inconsistent with our understanding of time as a strictly monotonous series of moments (in a macroscopic world). Brahe's cycloids do not facilitate a simple merging of geometric and temporal information, unlike Kepler's ellipses, which do. Remarkably, the church opposed the Copernicus/Kepler planetary theory for a similar reason: the inconsistency with existing other 'knowledge.' According to religious dogma, hell is in the middle of the Earth, and heaven is around the Earth. Where are all these if the Earth is not the center of the world?

Kepler's theory is preferable for its ability to deal with more affordances of nature. To what extent this might be the case is also evident from the fact that a program can implement Kepler and Brahe's theories today. Both approaches can be equally suitable for the computer (for simplicity, we assume that within some margin, the two models can be equally accurate and that computational complexity is not an issue). For the computer, it does not matter which program it executes; for us, the theory we take is essential.

AI artifacts can also be viewed from the perspective of their potential for information merging. Deep Blue has access to the rules of chess and, more importantly, to numerous games played in the past. For chess playing, this can be satisfactory. This view may not be the case if chess is understood broadly as a metaphor for human behavior and thinking, as in Nabokov's book, "The Real Life of Sebastian Knife." Traditional AI solutions may not be able to combine meta-level and object-level information, let alone improve their problem-solving strategy in this way. A lack of ability to systematically aggregate different types of information could pose a serious obstacle to AI in the future.

In current technology, programming languages only support a primitive form of information merging, for example, by allowing data to be coerced between different types, such as integer and real. Our brain is more flexible and can combine information from disparate domains, including natural language, logic, and reasoning. Familiarity in those domains allows a human agent to detect whether a statement can be consistently interpreted across all domains in a split second.

Can AI technology meet the demands of human interpretation in all respects? Experience shows that current AI solutions are promising but also have limitations. What this means for the future has yet to be made clear. The brain can store an enormous amount of information about the world. However, memory size is less and less of a problem for computer technology. The source of the problem of having computers that can think human-like lies in the complexity of meaningful processing, as Artificial Intelligence has painfully experienced several times.

2.4   AI Versus IA

What tasks can be safely entrusted to the computer? How far can automation be pushed? Deep Blue performs at an unprecedented level but does not play chess as humans do. While Kasparov's strategy may be complicated for a layman to understand, Kasparov can certainly explain his moves in simple language. A similar explanation is probably not feasible for Deep Blue. This program is based on a statistical analysis of chess games, particularly those played by Kasparov in the past. The results of such an analysis are difficult to explain due to the complexity of the calculations used and the assumed differences between the mathematical and the naive meaning of probability.

The computer can be helpful for tasks that suit automation. Still, its use is less evident for problems that require genuine intelligence. But even in those cases, the computer can be beneficial, not by replacing human activity, but by enhancing it. An example of such an application is the computer mouse. Calculations by the mouse software are hidden, yet their meaning, such as cursor positioning and selection of text fragments, can be evident and valuable to the human agent.

The above examples already illustrate the differences between two main approaches that computer science has seen since the 1950s, namely Artificial Intelligence (AI) and Intelligence Augmentation (IA). The goal of AI is to replace human agents with autonomous systems. On the other hand, IA strives to improve human intelligence through using computers. In environments where the input information and the goal set for the system are clear, AI has better chances. When human experience is essential, IA offers better opportunities.

Efficient human-computer interfacing, necessary for IA and, to some extent, also for AI, requires a model of information processing that respects the properties of meaningful interpretation. The computer must grasp the meaning of human communication to a certain extent, and vice versa; the computer's output must be comprehensible to the human agent.

Program intelligence can be increased by considering information from different knowledge domains and viewing the input from various angles. For example, the quality of language processing can be increased by integrating knowledge from different domains, such as logic and reasoning. It is doubtful whether this method of qualitative improvement can be achieved on an ad hoc basis. The complexity of the human intellect suggests that using a theory of interpretation is inevitable, which is an essential point, especially as current AI seems to ignore the need for a theoretical basis for meaningful processing. An example illustrating the possible consequences of this is the following.

Suppose you are unwell and consult a search engine with your complaints. The program recommends a medicine that you are not familiar with. Would you take the

medicine or prefer to consult your doctor first?[11] It is not unlikely that you will use
the drug suggested. For example, when the program can explain how it came to its
response and convince you in this way. How can this be achieved? What are the
conditions for a meaningful explanation? Since it is about the computer, we assume
that typical human characteristics such as reliability, persuasiveness, and emotional
expression do not play a role.

A fundamental question of meaningful processing is which phenomenon the
stimulus stands for. Therefore, a prerequisite for efficient human-computer communication must be that the computer output is presented so the human user can
understand it just as easily as any other stimuli. This condition is fulfilled by natural
language communication (which may also explain why it is called 'natural') but not
by most computer systems. Computer output generated by a program can be long
and difficult to comprehend. This shortcoming can be improved by eliminating parts
irrelevant to the meaning of the calculation as a whole, that is, to the phenomenon it
represents.

In sum, meaning-preserving summarization can be essential for modeling interpretation as a computational process. Human-computer communication is marginally important to Deep Blue. Its input is limited to chessboard configurations. As
opposed to chess playing, an efficient user interface is crucial for search systems.
Unless we uncritically accept the program's advice, we need a computer interface
that matches, at least to some degree, the ingenuity of the human intellect. Answers
like "the computer says no" are unlikely to be satisfactory.[12]

The importance of genuine artificial intelligence, i.e., computers that have learned
to think like humans, is broadly acknowledged. In the words of Nobel Prize winner
Gerard t'Hooft:

> If such intelligence can be constructed, it will quickly outsmart humans by a big margin.
> [. . .] The consequences are difficult to predict and even potentially dangerous — I don't fear
> that such a program will overthrow humanity or anything like that, but it might bring
> unparalleled power to those who are in possession of such a device.[13]

2.5 Breadth and Depth

We need to discuss one last topic to answer our original question about knowledge in
the Introduction, which concerns two characteristic features of memory search:
breadth and depth.

As mentioned earlier, information processing may be subject to temporary
conditions. But why would this be a problem? Although input processing can only

---

[11] A similar problem is described by Merredith Broussard (2023). A story from this book, "An AI
told me I had cancer," is published on https://www.wired.com/story/artificial-intelligence-cancer-
detection/

[12] https://nl.wikipedia.org/wiki/Little_Britain

[13] Nothing to fear from mistakes. Interview with Gerardus 't Hooft, Nature, 467, page S7 (2010).

be sustained for a limited time, it may be sufficient to allow obtained responses to recursively activate additional memory information and increase the breadth and depth of the search space. The question is: can the brain expand its search space to the entire memory in this way?

To be efficient, already visited memory elements must be remembered during a search. In addition to processing time limitations, the brain also has a limited ability to distinguish data. Search depth limitation is illustrated with the maximum look ahead in chess, which is about 2-5 moves for a layman and 15-25 for a world champion. Since there are 10^{50} configurations, which is way too high to consider all outcomes, a deeper look ahead can increase the chances of winning. Thinking ahead does not mean that every possible move is explored. Eliminating low-utility configurations, which reduce search breadth, can help increase search depth. AI recognizes this approach in the algorithm, known as branch-and-bound search.

The limitations of human processing were first studied by George Miller (1956). Miller considered the brain as a communication channel and examined the effect of input growth on the reliability of the output. In one of the experiments, listeners were asked to identify tones between 100 Hz and 8 kHz by numbering them. After each classification session, the test person was told the correct solution. This experiment showed that classification was precise when 2-3 tones were presented. The result was increasingly erroneous with four tones, and beyond 14 different tones, identification was very bad.

Donald Broadbent (1975), who repeated Miller's experiment 15 years later, concluded:

> Human processing is limited to handling a fixed number, say 7, independent units at one time. Each unit could nevertheless be divided into sub-units so that facts and actions of enormous complexity could be handled by calling up a fresh ensemble of sub-units at each stage when it became necessary.

2.5.1 Learning Ourselves

Using as much knowledge as possible increases the likelihood of environmentally significant responses to external stimuli. Ideally, all knowledge learned in the past should be used. We already know that this is not feasible. That is why we only ask what is needed to approximate that goal.

From a broader perspective, input processing is similar to looking at the outer world through a small window opened by the stimulus and telling what is there. The need to increase the quality of the response forces the brain to use maximum information about previous stimuli comparable to the current one. In light of the architectural limitations of memory access, this need can be met by remembering our earlier responses, thus allowing the brain to learn itself. Because those responses are ours, by remembering their information, our memory becomes an increasingly accurate representation of ourselves, i.e., what we are.

2.5.2   Why Can't We Tell Everything We Know?

A prerequisite for interpretation, and thus for access to stored knowledge, is the occurrence of a stimulus. There is also a second condition: only data matching the stimulus can be used to obtain a response. Therefore, we cannot tell knowledge (at least not by the current process) that does not respond to the stimulus within the time available for input processing. We can only tell the whole of what we know in the sense of tendencies and attitudes based on our knowledge of ourselves.

There is more to say about human processing limitations. Miller and Broadbent's experiments provide evidence for a phenomenon known as chunking, i.e., grouping connected items to be processed as single concepts. An example is our ability to split long numbers into shorter pieces that are easier to remember. From the perspective of our theory, chunking can be related to the brain's potential to process complex stimuli recursively.

Working memory capacity determines the available memory size for a single incarnation of interpretation as a process. In addition to the data storage size, the dependency between the processing events also sheds light on the organization of memory information. Knowledge is subject to and, in this sense, shaped by *architectural* constraints of cognition. We need a more refined model of human interpretation as a process to identify these limitations.

2.6   Towards a Model

The study of interpretation traditionally belongs to philosophy. However, except for Peircean sign theory, philosophy pays little attention to interpretation as a process, i.e., its dynamic properties, let alone to the computational aspects of human processing. Although Peirce's theory is not computational, his sign concept allows a process model for interpretation to be derived almost straightforwardly. On this sign-theoretical ground, we introduce our theory in the next chapter.

Chapter 3
Cognition Is Formed by Brain Processes

Abstract Experimental research on visual processing demonstrates that interaction is fundamental to brain activity. The ability of the brain/mind to transform static data from discrete input processing into dynamic responses is evidenced by the cognitive phenomenon known as apparent motion perception: series of static images are perceived as dynamic motion. The possibility of a single type of process follows from experimental research indicating simultaneous syntactic and semantic information processing by the brain/mind. How can information processing by the brain/mind be characterized? First, the input must be represented as a collection of data; second, the two types of input data (state and effect) must be recognized independently; and, third, in combination. Based on these considerations, an information processing model consisting of nine representation events can be defined.

3.1 A Miraculous Phenomenon

By now, we have a clear picture of the conditions for meaningful processing. But what is the cognitive reality of uniform information processing? What can be the events of that process? This chapter focuses on these questions. We illustrate our answers with evidence from cognitive research on visual perception.

Visual perception is known as an unusual phenomenon. Suppose you are looking at a picture hanging on the wall. Nothing else exists in your visual field; neither the image nor your eyes move. What happens is that after 4–6 s, you see nothing. Everything turns grey. The picture is still there, and your eyes are always open. It feels like you have gone blind. This experiment sounds so incredible that you want to check it out yourself. You select an object, stare at it motionless for a few seconds, and then . . . you will still see it. It does not fade, contrary to expectation. So, none of this is true.

Well, it is. The experiment's instructions must be followed precisely to prove that it happens. A crucial condition is that the image and the eyeball both are immobile. In particular, the second condition is challenging to sustain. Mechanically fixing the eye is painful; you should not try it yourself. A painless but technically more demanding solution is compensating eye movements by synchronously moving the image.

As early as the nineteenth century, scientists knew that eyeballs are never rigidly fixed and can be subject to vibrations. More than a century later, in the 1950s, Lorrin A. Riggs and Floyd Ratliff demonstrated that stabilized images lead to visual perception blurring and disappearance. But it was Alfred Yarbus who first found an explanation for this phenomenon a decade later. For his research into visual perception in the absence of eye movements, Yarbus devised an ingenious suction cup that can be attached to the eye painlessly. The results of his experiments were astonishing. Yarbus proved that a change triggers visual processing. It may be due to a change in the stimulus and, unexpectedly, to saccadic eye movements, which are tiny movements by the brain itself. The possibility of saccades explains the unusual phenomenon we discussed above, namely the disappearance of visual perception in the presence of a static stimulus, also known as retinal stabilization.

Yarbus' experiment shows that the occurrence of a *change* is a fundamental condition for sensory-level visual perception. Below, we will argue that the appearance of a change is also characteristic of higher organizational levels in the brain. To justify our claim, we focus on three increasingly evolved moments in visual information processing and analyze them from the perspective of change.

3.1.1   Photoreceptors

The first level we look at is when the incoming light signal activates photoreceptors in the retina. Light-sensitive neurons can occur in two states: active and passive. Active receptors respond to the light stimulus, discharge, and release an electro-chemical signal; passive receptors, which are charging, do not react. We know there is a change due to the incoming light signal; otherwise, there is no sensory processing. But is there also a change in the retinal output signal? If so, it may reinforce the hypothesis that lower- and higher-level processing follows the same pattern.

There is indeed a change, the source of which can be found in the difference between the output signal from the retina and the signal from the previously inactive and now-firing photoreceptors.

3.1.2   Optical Pathway

The retinal output signal is the input for next-level visual processing. Only light rays that follow the eye lens's orientation are in the retina's focus and are subject to processing. The retina has two classic types of photoreceptors, namely, rods and cones, which primarily contribute to nighttime and daytime vision, respectively. The output of these receptors is relayed through ganglion cells and, ultimately, the optical

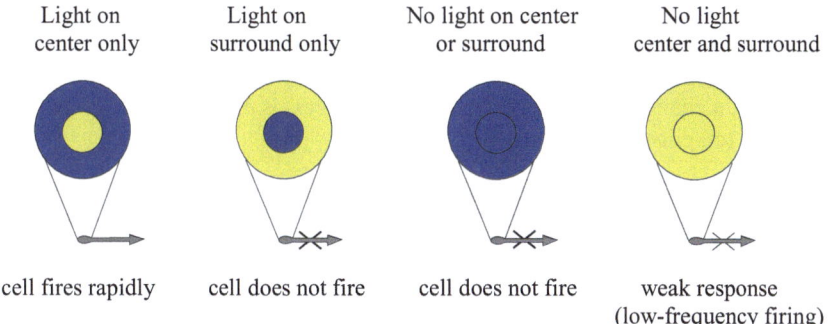

Light on center only — Light on surround only — No light on center or surround — No light center and surround

cell fires rapidly — cell does not fire — cell does not fire — weak response (low-frequency firing)

Fig. 3.1 On-center-off-surround-cell (based on https://en.wikipedia.org/wiki/Receptive_field). Cell fires at high intensity when light only falls on the center region (see leftmost diagram). In every other case, there is no or only a weak output signal

pathway.[1] In the late 1950s, Steve Kuffler discovered concentric receptive fields in the optical pathway. These cells can be organized in two ways: on-centered off-surround and off-centered on-surround.[2] See Fig. 3.1. The first type of cell fires rapidly when the light encloses the central (on-)region and is inhibited when the light encompasses the surrounding (off-)one. A light stimulus that spans both fields, the center and surround, produces little change in activity. Off-centered on-surround cells operate oppositely. The two types of cells can signify a change in illumination arising from stimulus edges (Gazzaniga et al., 2002). Accordingly, they transform a change in their input into a change in their output signal. The information conveyed by the output signal, input for higher-level interpretation, often specifies the edges of objects. The disappearance of a signal, which is a change, albeit in a negative sense, conveys no information. After all, no difference can be made between the lack of a signal and the inability of inactive receptors that (re)charge.

3.1.3 Visual Cortex

The signal that the 'on-off' cells produce is input for processing in the visual cortex. Based on Kuffler's results, in the 1960s, David Hubel and Thorsten Wiesel investigated information processing in the primary visual cortex by analyzing the activity of individual brain cells in cats. They have found that a change in the illuminating light, not the light qualities themselves, elicits a response from cortical neurons. Unlike the cells of the optical pathway, cortical neurons respond to edges. Hubel and

---

[1]Cf. the LGN or lateral geniculate nucleus.

[2]https://www.researchgate.net/publication/241258132_A_robust_sub-pixel_edge_detection_ method of_infrared_image_based_on_tremor-based_retinal_receptive_field_model

Wiesel's research demonstrated that information by visual cues in memory is expressed by a topographic map of specific edge orientations, such as horizontal (−), left-diagonal (\), vertical (|), etc., justifying that also at high-level information processing, a change in the input (cf. appearing edges) is represented by a change in the output (cf. edge orientations).

3.1.4   Interrelated Representations

Interpretation by the photoreceptors, the on-off cells of the optical pathway, and the cortical neurons are related. The signal by the photoreceptors is transformed into a signal by the optical pathway and, in turn, into an edge orientation by cortical neurons. The other way around, the output of the photoreceptors can only appear as an edge orientation through the mediation of the signal by the optical pathway. The three signals represent the input visual phenomenon from different perspectives: the first deals with the qualities of the input itself (e.g., intensity, color, hue, etc.), the second with the relations of the input to other qualities (see center/surround cells), and the third with the input meaning (cf. edge orientations). In Chap. 6, we will argue that an analogous ternary classification applies to all phenomena, not just visual ones.

3.2   Interaction and Change

The above explanation of visual processing sounds clear, yet there is a problem. Previously, we argued that an interaction between the stimulus and the brain as an interpreting system is necessary for information processing. According to Yarbus' experiments, such a prerequisite must be the occurrence of a change, not an interaction. Are we mistaken?

Fortunately, we are not. It can be shown that all interaction involves a change. To justify this, we need to rethink the concept of interaction. In essence, such an event presupposes the existence of qualities that are, in principle, independent. Regarded as qualities, the brain (including the senses) and the external stimulus fulfill this condition. Conversely, qualities that are not independent, hence not different, cannot interact; their apparent 'co-occurrence' does not qualify for a phenomenon. Consider, for example, painting over the wall of a certain color with an identical color. Since the two colors, old and new, are not independent,[3] you will not experience a change.

Since interacting qualities affect each other, which must be the case, and otherwise, there would be no change, they must be compatible in some sense, in addition

---

[3]Qualities that, as collections, are in a subset relation with one another are not independent.

Fig. 3.2 Mosquito bites

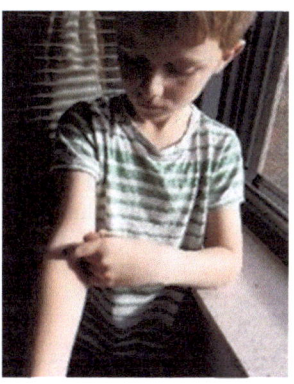

to being different. An example is the interaction between the sound of train noise and the eardrum. It needs no explanation that, as qualities, they are different. That they are compatible stems from the ability of sound to activate the eardrum and the eardrum to respond to sound vibrations.

In summary, interacting properties must be different but also have compatible properties. Accordingly, every interaction involves a change, defined by the difference between the interacting qualities compared to what they have in common. The dependency between interaction and change has significant consequences. Meaningful representation demands that all events are an interaction. It follows that all events, including those of the interpretation process, involve a change. If this last condition were not the case, input processing would suffer from information loss and be incomplete. Also, the meaning of the stimulus must be explainable in terms of the change involved in the input interaction. Because interactions are dynamic, the input meaning must also be dynamic. It thus cannot be expressed solely by a static value such as a fact.

3.2.1   Sample Analysis

To gain insight into the process meaning of interpretation, we take an example with which the reader may have more experience than trains and level crossings, namely the phenomenon of mosquito bites. For our analysis, it is sufficient to know that mosquitoes penetrate the skin and inject saliva, i.e., anticoagulants that cause an inflammatory reaction, swelling, and itching in the deeper skin layer. Below, we assume that the input is defined by the sensation of piercing the skin and irritation from saliva injection and the response by a motor action of scratching the site of irritation. See Fig. 3.2.

Interpretation is a transformational process, from input interaction to meaningful response. Much is known about the underlying neurophysiological phenomena, except their overall goal. The question of why neurons in the brain function the way they do has yet to be satisfactorily answered. One reason is the great distance between neuron-level and high-level human processing. The model of this book, which attempts to provide an answer based on a theory of meaning, comes in handy. To elaborate on this further, we need to explain the events of the interpretation process. However, this goal is too ambitious to achieve right away. Therefore, we begin with interpretation as a series of increasingly complex stages of input expression, which we then expand into a series of events. We then delve into the function of these events to ultimately reveal the rationale behind meaningful information processing.

3.3   Three Stages of Information Processing

The transformation from stimulus to response can be split into three stages. See Fig. 3.3. In the first stage (1), the piercing and irritation by saliva injection are transformed based on sensory memory information into a 'feeling' of itching, representing the stimulus as a collection of qualities. There is no recognition yet of the interaction between saliva injection and the skin due to a mosquito bite, and there is no reaction either.[4] In the next stage (2), the 'feeling' is used to find memory information about similar feelings in the past. In other words, it is transformed from quality into a relation with stored information. We are done if the entire stimulus can be recognized in this way. Otherwise, additional data, possibly related to the input, is required.

An example is the location of the piercing (in the soft inside part of the elbow, or simply the elbow) and the properties of the 'feeling' of itching. In stage (3), the resulting data is checked for completeness and consistency, whether it explains all input qualities individually and the relationships between them. If this holds, the 'feeling' is transformed into a more developed relation that involves a response to the stimulus in addition to input and memory information already available. An example is the thought of 'incipient itching due to mosquito bites in the elbow' or the motor action 'scratching the elbow.'

The three stages are related to each other in two ways. Through a relation of *dependence*, since information by (3) presupposes data by (2), which in turn needs input from (1). For example, scratching the elbow assumes information about the location of mosquito bites and, in turn, about a 'feeling' of piercing. Second, through a relation of *subservience*, since (1) to arise as a meaning needs (2), which in turn needs (3). For example, the 'feeling' of irritation (1) can arise as a response in (3) through the mediation of (2). All three stages involve the meaning of interaction: in stage (1), between stimulus and sensory memory, representing the stimulus as an occurrence of independent qualities and, thus, as potential meaning; in stage (2), between input and memory, representing their relation as factual meaning; in stage (3), in the combination of input and memory information with a response,

---

[4]This also explains the use of apostrophes in the term 'feeling.'

stage (3)	'feeling' of itching	—	in the elbow — scratching the elbow
stage (2)	'feeling' of itching	—	in the elbow
stage (1)	'feeling' of itching		
	(saliva injection)		

Fig. 3.3 Increasingly complex meanings according to the three-stage model

representing the 'feeling' as actual meaning. Since the stages of processing are experienced, they must be phenomena. In summary, interpretation involves three stages, or types of phenomena, embedded within each other. The question of how the creation of their increasingly evolved input expressions can be described as a procedure asks for a more refined analysis of meaningful processing.

Before we go any further, we need to say a word about the difference between 'involve' and 'contain' and why we prefer the former over the latter. The main difference is that the object of 'contain' is an object, while the object of 'involve' is an action. The first is static, and the second is dynamic information. 'Contain' is to hold something within a limited area or volume. 'Involve' is to add or have as an integral part of a process. Such as an appearing property. Such as meaning. Because of this difference and the dynamic nature of the process of interpretation, we prefer the verb 'involve' to emphasize that phenomena mean more than a simple relationship. Also, instead of 'make' and 'generate,' we use the verb 'obtain' or 'create' to indicate that phenomena and meaning do not result from a mechanical process. Involved and involver are inextricably linked. Analyzing dynamic interpretation into static relations deprives it of its authentic meaning.

3.4 Qualities and Qualia

Earlier, we argued that all events, including the events of the process of interpretation, must be interactions between independent qualities. Because every process is goal-oriented, every event must arise from previous events (except the input), and the set of events must define a hierarchical order. In terms of meaningful processing, this means that its events must be able to function as a quality in addition to their meaning as a result. What does cognitive theory say about qualities?

According to a theory of perception (Harnad, 1987), external qualities are internally represented as sense data, or 'feeling,' by *qualia*.[5] The concept of a quale (singular for qualia) derives from the Latin adjective 'qualis' meaning "of what sort" or "of what kind" in a specific instance, i.e., individual instances of conscious

---

[5] 'Sense data' is an epistemic, and 'qualia' is a phenomenological concept. The first is about whatever fact results from sense experience; the second is about what it is like, i.e., the properties of sense data in our observation. We assume the two concepts are interrelated and treat them the same. C.S. Peirce introduced the term quale in philosophy in 1866. https://plato.stanford.edu/entries/qualia

experiences, such as "what it is like to taste a specific apple, this particular apple now."[6]

The cognitive concept of qualia is founded on categorical perception or organisms' potential for sorting the world's objects into categories. This potential was first observed in color perception and the perception of speech bounds, but it has since been found in various domains. Equal-sized physical differences between stimuli are perceived as larger or smaller depending on whether the stimuli fall into the same or different categories. For example, a pair of greens of different shades look more like one another than a shade of yellow (which may be no more different from one of the greens than the other green is, e.g., in wavelength), and this difference is one of quality. Also, the concept of a threshold, a term we have often referred to, is rooted in the idea of categories. In a broader context, using categories is also practical from a computational stance, as it allows for error-correcting coding. In the 1940s, digital data representation, and thus the computer as we know it today, was introduced as an alternative to analog signal processing for the same reason.

In summary, we assume that qualia characterize all levels of interpretation, not only sensory perception. For example, thoughts arise from qualia and are qualia themselves. Analogous to quality, a quale can designate a single quale and a collection of qualia ambiguously.

3.5   Discrete Processing

In addition to using qualia, discrete processing is another alleged characteristic property of the brain. External stimuli are sampled in a collection of qualia, thus represented as static data. By keeping the input invariant, as long as sensory activation permits, such a sample enables memory information to be explored systematically.

All clear, but there is a problem. If interactions are dynamic, how can the meaning of the stimulus and, thus, the phenomenon signified be represented only by static information? If interpretation suffers from information loss due to discrete processing, the response to the stimulus may not be helpful. We are faced with a severe dilemma. Either the assumption of discrete processing is incorrect, or static data can still be enough to represent the meaning of dynamic phenomena. But how?

3.5.1   Cognitive Reality

It may come as a surprise, but static data may suffice. Evidence can be found in apparent motion perception. In this phenomenon, a series of steady images are

---

[6]https://en.wikipedia.org/wiki/Qualia

Fig. 3.4 From static images to dynamic motion perception

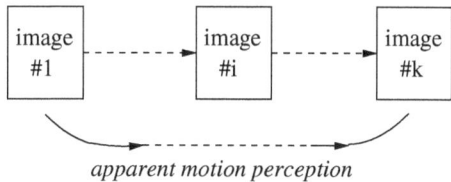

apparent motion perception

presented. Although each image can be meaningful, considered in itself, combined, they can be interpreted as parameters in the experience of dynamic motion. However, there are a few prerequisites for this experience. Each image must be perceived on its own (1), changes between successive images must be identified (2), and how these changes may contribute to a sense of motion must be explained (3).

Apparent motion perception is an intricate phenomenon. In addition to the conditions listed above, many more conditions must be met for a sequence of images to create the illusion of a smoothly moving scene, known as beta motion. The images must be projected with a specific frequency, approximately 10 to 12 images per second. It must be dark for some time between the images, and the difference between successive images should not be too significant, to mention only a few. See Fig. 3.4.

Besides motion perception, discrete processing can be satisfactory for dealing with dynamic phenomena in general. Juggling is one of them.[7] By practicing with flying objects, such as rings, one can learn to recognize 'key' configurations for catching and throwing rings (3), notice changes in their positions (2), and observe their kinetic properties (1). That apparent motion perception and juggling can be distinguished into three different kinds of experience is no coincidence. In Chap. 5, we will argue that all phenomena can be characterized similarly.

3.5.2 State and Effect

How does the interpretation process represent the dynamic character of the input interaction? This exciting question, when answered, will ultimately lead to a deeper understanding of meaningful processing. To this end, we need to rethink the concept of quality. This analysis shows that the ambiguity of this notion, the key to the input representation, is also a primary source of interpretation complexity.

Let us return to our initial question about. Interpretation has three sites: stimulus, memory, and memory response. Previously, we assumed that a collection of qualia represented the stimulus. Memory[8] and, so, the memory response triggered by the input must be a collection whose elements are themselves collections of qualia—in

---

[7] https://www.wikihow.com/Juggle-Three-Balls, https://www.workshopjongleren.nl/leren-jongleren

[8] The memory information is stored data about phenomena observed in the past.

other words, a collection of collections of qualia. A similar concept in mathematics is known as a multi-set, a set whose elements are a set. What makes sets particularly interesting is that they provide access to their elements. Still, a set can also be treated as a single value. Examples are input and memory, collections of qualities that function as a single value in the interaction that produces the memory response as a set. However, treating collections as a single quality is controversial in light of a process view of interpretation. How can information from *two* independent qualities, stimulus, and memory, be represented by a *single* quality, the memory response?

We can find an answer to the above question in the interaction properties discussed earlier, the occurrence of independent qualities and the arising change, and in the condition that, to be meaningful, those properties must be represented by information processing. Let us see what all these imply.

Since the stimulus interacts with (sensory) memory, they must be independent. Accordingly, memory response triggered by the stimulus must allow for the interacting qualities to be distinguished into two types: shared and different. At least two response elements are needed to represent both types of qualities. The assumption of memory as a homogenous collection has to be revised.

Although memory can be considered a collection, its elements must be independent.[9] This property is the key to a memory representation of the input interaction. Qualia shared by and different between memory response elements can represent the two types of qualities of the stimulus. Moreover, a relation between their data can express the arising change. Which memory elements can be compared? Those that are related, i.e., linked with each other. In matching mode processing we assume that there are responding elements that are directly connected in memory. In analysis mode, such a connection is not available and the task is to connect agreement sense responding elements transitively, via complementary data. If a link can be found in this way, the arising dependency can be remembered by introducing a direct connection, which changes the network structure of the memory. See Fig. 3.5.

In summary, by comparing a pair of response elements, the interaction between stimulus and memory can be represented by qualities shared by and different. The emergent collections of qualities, which we call *state* and *effect*, are independent due to the independence that characterizes stimulus and memory, and the independence between memory elements. In this way, discrete processing lays the ground for the duality of the input offered for information processing.

Why state and effect? Since shared qualities are involved in both the stimulus and the memory response, they are unaffected by the interaction. Those qualities were there (that is, represented in memory) before the interaction and are still there (in the stimulus), so they qualify for a *state*. We are particularly interested in a representation of stimulus qualities that differ from those of the memory response. The idea is that they can be defined by qualities that were not there (not represented in memory) before the interaction but are there now (in the stimulus). Such qualities can be seen

---

[9]Only if the stimulus is unknown is it offered for memorization. Accordingly, memory can be understood as a set (elements of a set are different by definition).

Fig. 3.5 Matching mode (left) and analysis mode processing (right). Black/white bullet = high/low activity memory response; solid/dashed edge = existing/new link. Responding memory information connected by an edge can be subject to comparison. Qualia that are shared and those that are different can be used to define a state and an effect

as an *effect*. From the above definition of state and effect, it follows that, in principle, a single response element is sufficient to represent a state, in contrast to an effect, which needs a pair of memory elements to define.[10] We can conclude that a state can exist on its own, but an effect always presupposes the existence of a state.

The two types of qualia inform how the state (see shared qualia) is changed by the effect (see different qualia) into a new state. Something we are familiar with has changed, and in which sense. The analogy with apparent motion perception, consecutive images contributing to a change, and finally, an experience of motion as quality should be clear.

What about qualities that were there before the interaction (in memory) but now are *not* there (in the stimulus), that is, disappear? As explained about the retinal output signal, such qualia do not convey information about an effect and, therefore, not about a phenomenon.

3.6 Two Modes of Processing

The concepts of state and effect are fundamental. Still, there is more to say about the input for information processing. The qualities offered for information processing result from the brain's ability for selective attention, a phenomenon known from cognitive theory. Those selected can activate memory. As a result, memory can respond in the sense of agreement and possibility. Regarding information processing, data from the former are central to the interpreting system. In contrast, data from the latter can complement information from the former. As a result, we can distinguish the collection of input state and effect qualia into focus and complementary.

Interpretation aims to establish a relation between state and effect qualia that are in focus. We are done if the stimulus can be *immediately* recognized in this way; this

---

[10] A pair of connected memory elements is also needed to represent a state-effect relationship.

is what we earlier called matching mode processing. If immediate recognition is not possible, interpretation attempts to recognize the input using complementary information looked up in a *dynamic* process. We called it analysis mode processing. One of the questions for this latter mode of processing is how complementary qualia can contribute to input recognition. Complementary information refers to phenomena observed in the past, so its qualia must have two types: state and effect. Complementary states and effects must be compatible with the input state and input effect in focus to function as complementary data. They must also be mutually compatible since they refer to the same phenomenon. From the perspective of input processing, their information is not independent since they arise in response to the (same) external trigger. They are like two sides of a coin. Here is an example.

Suppose you look at a landscape that reminds you of a painting you saw before. That painting and the way of seeing it in the past (complementary state and effect) must be compatible[11] with the qualities of the scenery and the way of watching it now (state and effect in focus). Since the complementary state (painting) and effect (seeing) refer to a past observed phenomenon, they are trivially compatible. They synonymously complete the meaning of the input observation.

3.6.1   Dynamic Memory

We mentioned that, upon termination, analysis mode processing may introduce a new link between memory elements. What is the cognitive reality of such adaptive memory?

In the Introduction, we discussed memory's capacity to store knowledge or learned information in a network of neurons. One of the brain's fundamental powers is its ability to quickly adapt to environmental changes and thus develop new habitual knowledge. The brain primarily accomplishes this by enabling changes in the connectivity between individual nerve cells. The connecting elements between neurons, also called synapses, can be modulated in their strength by various mechanisms through synaptic plasticity.

According to experiments by Florian Engert and Tobias Bonhoeffer in 1999, new spines appear between neurons after induction of long-lasting (but not short-lasting) functional enhancement of synapses. The new synaptic connections allow the brain to adapt to the changing world. Although new spines appear roughly 30 min after stimulation, a connection is operational only after 16 h.[12] This potential of the brain

---

[11] The landscape can be similar to the current one, and watching in the past is analogous to seeing the scenery now.

[12] Experimental research by Yang et al. (2024) provides evidence that supports the hypothesized mechanism behind the organization of network memory. The study suggests that when at least 15% of hippocampal neurons are simultaneously activated during the processing of the input stimulus, their information will be retained (in working memory) and remembered during sleep (through establishing a link between responding neurons).

is reminiscent of folk wisdom, which says that in case of a complex problem, you should sleep on it before making a decision. It is tempting to conclude that, due to the brain's reorganized connections, you will see yesterday's problem as a solution the next day.

The most exciting thing about Engert and Bonhoeffer's experiment is that it demonstrates the brain's ability to reorganize its memory in response to external stimuli and transform itself, in addition to converting input data to output data, which a computer can do.

In summary, memory access can operate in three modes: first, associatively or 'by content'; second, via a network of links or 'by address' in a sense; and third, through habit formation or by reorganizing memory. The first two modes involve the element of a transformation, albeit a less developed one, such as the activation of neurons and comparing their data. Still, the adaptive memory experiment alone provides evidence for transformation by the brain, which is undoubtedly related to meaningful processing.

3.7 Toward a Process Model

Much has been said about interpretation's properties; now it is time to explain it as a process. It's tempting to think of the relationship between stimulus and response as analogous to the action-reaction dependency in Newtonian mechanics, but that's wrong in many ways.

Interpretation is not a mechanical process, but we already know that very well. Unlike the Newtonian world, the relation between stimulus and response does not arise instantaneously. Stimuli can be complex, so we have reason to believe that interpretation consists of a series of events, not just a single event. Those events which are governed by a goal, namely to give a response to the stimulus, define a process. Since they are determined by the interpreting system, at least partly, they must appear in its experience.

Below, we introduce interpretation as a series of events not based on neurophysiological evidence but derived from assumed properties of meaningful processing. Although the resulting model is conceptual, we will show that its events conform to the characteristics of human processing. We anticipate this conclusion and deal with their results as meaningful representations. The interaction between stimulus and interpreting system can be viewed from the perspective of either one of them. Here, we take the side of the interpreting system, e.g., the brain. Since we can only speak of a response to the stimulus when interpretation terminates, its events, except the last one, are only becoming meaningful and, thus, are incomplete.

3.7.1 Input Qualia

Information processing begins with the determination of the input qualia. We have stimulus and memory and distinguish memory response into two types: agreement- and possibility-sense. The two types of response allow the relation between stimulus and memory to be defined as a state and effect in four different ways. The resulting relations define the types of input qualia: state in focus, effect in focus, complementary state, and complementary effect.

The event defining the individual input qualia is called *input*; the arising collection of qualia is referred to as 'primordial soup' or <u>*input*</u> for short.[13] We use the convention that processing events and emerging input representations are given by symbols in italics and underlined italics, respectively. Recall that a quale can ambiguously denote a single quale and a collection of qualia.

That said, we finally have all the ingredients for our model that we will consistently call the *process model*. Now, let us see what the events of meaningful processing are. In light of the purpose of the interpretation, their definition can be simple. However, this is only apparently true. We must remember that interpretation is qualitatively more complex than a sequence of events. The question of how the process model relates to authentic interpretation still has to be explained.

Below, we begin with analysis mode processing, followed by a model of matching mode processing. The processing events are looked at from three perspectives: the point of view of interaction, input representation, and logic.

3.7.2 Processing Events

To determine the input meaning, the four types of qualia in the input primordial soup (<u>*input*</u>) must be first individually recognized. To this end, the input collection of qualia has to be conceptually transformed into a set. This event, which we call *sorting*, provides access to the individual input qualia (recall that a collection, unlike a set, does not have this ability). Following the goal set for interpretation, namely determining a relation between the input state and effect in focus, their qualia must first be identified in the input. We call the result of this event <u>*state-in-input*</u> and <u>*effect-in-input*</u>. See Fig. 3.6. Complementary state and effect qualia, which are not in focus and respond to the stimulus in the sense of possibility, are collectively represented by <u>*context*</u>. All representations so far refer to the input primordial soup. Establishing the meaning of the input state and effect, which are in focus, asks for their qualia to be represented separately. The result of this event, called *abstraction*, is denoted by <u>*state*</u> and <u>*effect*</u>.

---

[13]Note the difference between stimulus and input. The first is external to the interpreting system, while the second is internal.

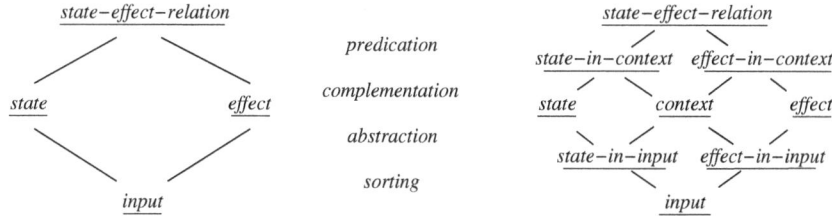

Fig. 3.6   The events of matching mode (left) and analysis mode processing (right)

Analysis mode processing assumes that only a subset of the stimulus can be recognized in an agreement sense. Processing the entire stimulus requires complementary information. That data must be compatible with the input qualia recognized in an agreement sense, i.e., that are in focus. The two types of information, focus and complementary, are combined in the event of *complementation*; the resulting input representations are called *state-in-context* and *effect-in-context*.

When the meanings of the input state and effect in context are consistent and compatible, information processing can achieve its goal by establishing a relationship between them. To this end, the input state and effect in context, as subject and predicate, are merged into a single relation. This event is called *predication*; its result is denoted by *state-effect-relation*. Consistent information processing demands that complementary data completing the abstract state must also be compatible with the abstract effect.

The above events define analysis mode information processing. The less complex matching mode processing does not use context information. The entire collection of state and effect qualia in focus can be immediately recognized, and *sorting* and *complementation* reduce to an empty operation. A characteristic feature of the events in both processes is that they all refer to the input collection and thus represent the input qualia. The relationship of the events with the three-stage model is illustrated in Fig. 3.7.

3.7.3   Independence

All this is promising, but we are on the right track only if we can demonstrate that each event is an interaction between independent qualities. This condition holds for the *input* event and is easy to see. Stimulus and memory interact, so they must be independent, and memory in the sense of agreement and possibility are independent because of their different activation intensity.[14] In sum, the four types of input qualia must be independent. *Sorting* and *abstraction*, the events that identify the different input qualities in themselves, are interactions, albeit in a negative sense, as

---

[14]Different activation intensities indicate different, and therefore independent, memory elements.

Fig. 3.7 The process model (left) and the three interpretation stages (right). In the diamond on the left, the lowest quadrant implements stage (1), the quadrants in the middle stage (2), and the topmost quadrant stage (3)

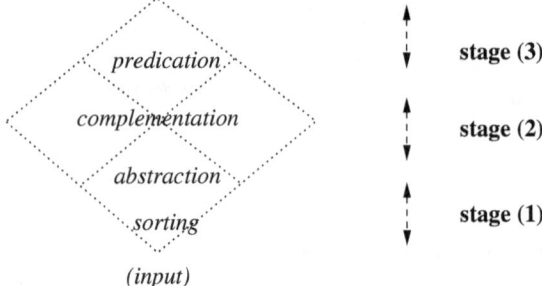

(input)

separation. Due to the independence of state and effect, *complementation* meets the condition imposed on interaction (<u>*state*</u> is completed by effect type and <u>*effect*</u> by state type information). This property also applies to *predication*, which combines the independent input state and input effect in context into a single relationship. The events of the process model, as interactions, are depicted in Fig. 3.8.

3.7.4   Input Representations

Since all events are interactions, the process model respects the dynamic nature of the input phenomenon. The question is whether those events can be meaningful or are just formal, like computer calculations. It is tempting to say that they are meaningful. However, we can only justify this claim based on a theory of meaning, which we have yet to introduce. We can achieve that goal, at least in part, by illustrating the naive meaning of the processing events. But even that goal is beyond our reach. The problem is that, from the perspective of input processing, those events (except the last one) are only becoming meaningful and thus have yet no meaning. We can only know what happened and respond to the stimulus accordingly when the process terminates. That said, the best we can do is to set out the individual events and leave it to the reader to merge their explanations into a single meaning. It is like showing images with long pauses between them and asking the audience to imagine the meaning of their entire sequence. Evidence shows that apparent motion perception also occurs in this case, but the emergent experience is less intense. As far as we are concerned, what matters is that motion experience can happen also in that case.

Below, we offer an analysis of the events of the process model. We illustrate their meaning with our example of mosquito bites.

In the *input* event, data for information processing is defined based on the response by mechanoreceptors[15] and chemical receptors of the skin. Their signal,

---

[15] https://www.sciencedirect.com/topics/biochemistry-genetics-and-molecular-biology/pressure-sensation

Fig. 3.8 Interaction events in the process model. A horizontal edge represents a single event

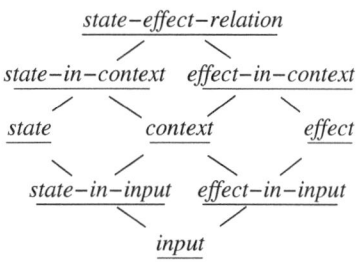

transmitted to the brain via somatosensory pathways,[16] activates memory. The 'feeling' of mosquito saliva can trigger a high-intensity memory response, e.g., about injection by (any) insect, impression by skin bite, and a low-intensity memory response about anopheline (the protein secreted in mosquito saliva), a degree of elasticity of the skin, as well as information about mosquito bites, its location, such as the elbow, to name a few. Qualia from the stimulus and memory response form the basis for expressing the input state and effect meanings in the sense of agreement and possibility.

In *sorting* and *abstraction*, the qualia of irritation by saliva are interpreted as an injection affecting a skin-like object (*effect*). The qualia of the skin are recognized as a state, which can be subject to any injection-like effect (*state*). In the next event, *complementation*, those abstract meanings are completed with contextual information. Injection, as an effect, is completed with information about mosquitoes (*effect-in-context*), and skin, as a state, with data about the injection's site, such as the soft side of the elbow (*state-in-context*). Finally, in *predication*, the arising state and effect meanings are verified for compatibility, whether they have common qualia, and for completeness, whether they explain the input as a whole. If so, the input state and effect meanings in context are combined into a single relation, which is a proposition. For example, the statement that your skin on the elbow is affected by a 'mosquito bite' (*state-effect-relation*).

The final response may arise as a thought and evoke a motor reaction, such as scratching or hitting the elbow. Since complementation can explain the input from a single perspective, interpretation must be hypothetical. Via communication, the emerging response can be reinforced but also refuted. If it was not a mosquito bite that caused the input sensation, but you (girl)friend tickled your arm, hitting the elbow can have unpleasant consequences.

When interpretation reaches its goal, it is done. What can be remembered from the processing events? Since the number of events can be considerable, due to the possibility for recursive input processing and for finding suitable complementary data, it is unlikely that remembering their information, some of which do not contribute to the input meaning, would be practical. We assume only the final input representation is offered for memorization and learning. Not as authentic

---

[16]Lower-level sensory phenomena also arise through interpretation, but their processes are beyond our interest.

meaning, but only as a *relation*. Such relations form the basis for memory as a network organization.

3.7.5 Remarkable Properties

The representation adopted by the process model has a few notable properties. One of them is the dual meaning of context information (*context*). This duality plays a vital role in analysis mode processing when the collection of input qualia cannot be immediately recognized. To that end, the process uses complementary data, which has two types: state and effect. Meaningful representation demands that, in *complementation*, the abstract state (*state*) is completed with information about an effect. Because complementary memory information responds to the stimulus in the sense of possibility, such an effect must indicate a property that is not characteristic of the input as a whole. Its qualia must be compatible with a subset of yet unrecognized input qualia and, in this way, complete the meaning of the input state.

The *complementation* of the abstract effect proceeds analogously. Like the abstract state, the abstract effect (*effect*) is defined by a collection of qualia. It can be similarly completed with information about a property. Because an effect always presupposes the existence of a state, the abstract effect can also be supplemented with information about what is affected, which information must be of state type. The goal of interpretation, namely determining the meaning of the entire input in a single proposition, demands that, in *complementation*, the abstract state and effect are consistently completed with memory data. Due to its synonymous meanings as a state and effect, the *context* can satisfy this condition.

In summary, *context* information functions in three ways: as a quality, compatible with the qualia of the abstract state and effect; as a relation, completing the abstract input expressions with complementary data; and, as synonymous meanings from some perspective, like the two sides of a coin.

In the running example, the abstract state is defined by the qualia of skin (*state*). We used complementary information indicating the site of interaction (*in-the-elbow*) as an appearing property and, thus, as an effect. That property, this time, used as a state (*elbow*), completes the effect (*effect*) with information about the object of injection. Note that the complementary effect (*in-the-elbow*) can also furnish the abstract effect, as a state, with information about how to give an injection. The terms *elbow* and *in-the-elbow* synonymously refer to the site affected by the stimulus.

Another interesting property of the process model's representation is due to effect-type data. The state involved by an effect allows the latter's meaning to be characterized in a rule-like manner: 'For this type of state, the effect has this meaning.'[17] An example is an injection, which can be given differently depending on the substance injected, such as insect saliva or liquid drugs. Still, liquid medicine

---

[17] An analogous language syntax characterization is called a verb's argument structure.

can be injected the same way as other kinds of substances, which justifies a rule-like definition of injection as an effect. By completing the meaning of the abstract effect with information about an involved state, *complementation* may reveal how the input effect is conventionally known (*effect-in-context*). Effect-type data can also be interesting for another reason. Since different effects can refer to an identical involved state, effects as qualia collections can be compared. Examples are mosquito bites, injection of drugs, and blood transfusion. As effects on a skin-like object (see involved state), they can be compared based on the pressure applied or amount of liquid material used, to name a few, and from that perspective characterized as different measures of an injection effect.

3.7.6 *Processing Capacity*

How does our process model relate to experimental research on meaningful processing? In this regard, we recall the influential result by Miller and Broadbent, which was mentioned earlier. Their experiment suggests that human processing is limited to simultaneously handling seven independent units in working memory.

The process model has nine interpretation moments. The first one, *input*, amounts to a definition of qualia offered for information processing. The emerging data can reside outside the working memory,[18] presumably in the thalamic area (Chai et al. 2020). The last one, *state-effect-relation*, can also be stored outside the working memory, e.g., in brain areas dedicated to motor control (nested phenomena, processed recursively, are represented as qualia of the embedding phenomenon without separately storing their information). In summary, seven interpretation moments must be remembered during input processing,[19] which fits well with Miller and Broadbent's conjecture about working memory span.

3.8 Meaningful Processing

We have made some progress toward our goal of a human-like thinking computer. Yet, we cannot sit back in our chair. Without a theory of meaning, we cannot prove that our model respects the properties of human interpretation. Fixing this problem is demanding and must wait until the next chapter. However, we can confirm that the process model can be meaningful, at least in a naive sense, by showing that its interpretation moments are compatible with concepts known from human knowledge. We will argue that such a link is available for propositional logic. In Chap. 5, we will suggest that natural language is no exception either.

---

[18] https://en.wikipedia.org/wiki/Working_memory

[19] In one incarnation of the process.

Propositional logic is a branch that studies the formation of complex propositions from more simple ones using logical operators. In English, words like "or," "and," and "not" are all such operators. The first serious study of propositional logic goes back to Aristotle. Medieval thinkers like Willem van Ockham have undertaken more work, particularly on logical operators. The next significant step forward came much later with George Boole's advent of symbolic logic.

The fact that a propositional-logical meaning can be attributed to interpretation follows from the goal set for the latter. To answer the question, "Why is this state affected by this effect?" requires a relationship between the input state and the effect to be determined, which also has a logical meaning. Following this line of thought, we argue that all moments of interpretation can be assigned a logical relation, not a full-fledged one, but one becoming meaningful as a proposition. Since those relations are not obtained through logical operators, the process model can be positioned as a 'naive' logic, a precursor of propositional logic (Sarbo et al., 2011).

The logical *not*-operator illustrates the difference between propositional and naive logic. This operator designates negation, e.g., if proposition p is false, $not(p)$ will be true. Unlike propositional logic, which is concerned with forming propositions and determining truth values, interpretation aims to establish a relationship between information in focus. Accordingly, the naive logical meaning of $not(p)$ denotes data that is not directly related to p and, therefore, complementary.

3.8.1 Logical Analysis

Information can be viewed from different angles, and the logical perspective is one of them. We ask what logical meaning the model's input representations entail. Due to its relational nature, those expressions fit perfectly with the concepts of propositional logic. In this logic, everything is a relation between propositions (unary relation included). Here is an example.

If a mosquito has bitten your skin on the elbow, you can conclude that the qualities of your skin and those of mosquito bites play a role. By doing this, you interpret the input interaction as a logical and-relation. The fact that the operands of a relation are identically treated as propositions implies that the distinction between state and effect can be ignored as to their logical meaning.

Below, we offer a logical analysis of the process model by looking at each event and asking what naive logical relation it entails. The result of this inquiry shows that interpretation, as a process, is logically complete. See also Fig. 3.9. Readers only interested in the main ideas can safely skip this part and proceed to Sect. 3.8.2.

The first event (*input*) represents the interaction between the stimulus and the interpreting system. Although the input state and effect are different, they are recognized as independent potential propositional meanings, having a truth value. Accordingly, the input (*input*) can be logically defined as a collection of

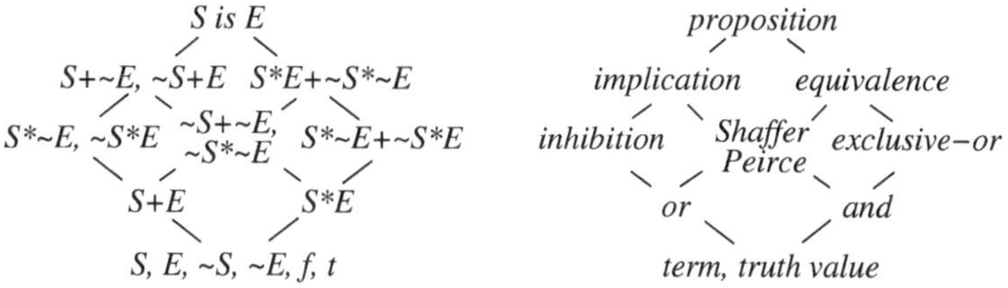

$$
\begin{array}{ccc}
 & S \text{ is } E & \\
\diagup & & \diagdown \\
S+{\sim}E, \; {\sim}S+E & & S*E+{\sim}S*{\sim}E \\
\diagup \quad \diagdown & & \diagup \quad \diagdown \\
S*{\sim}E, \; {\sim}S*E & \begin{array}{c}{\sim}S+{\sim}E, \\ {\sim}S*{\sim}E\end{array} & S*{\sim}E+{\sim}S*E \\
\diagdown \quad \diagup & & \diagdown \quad \diagup \\
S+E & & S*E \\
\diagdown & & \diagup \\
 & S, \, E, \, {\sim}S, \, {\sim}E, \, f, \, t &
\end{array}
$$

$$
\begin{array}{ccc}
 & proposition & \\
\diagup & & \diagdown \\
implication & & equivalence \\
\diagup \quad \diagdown & & \diagup \quad \diagdown \\
inhibition & \begin{array}{c}Shaffer \\ Peirce\end{array} & exclusive{-}or \\
\diagdown \quad \diagup & & \diagdown \quad \diagup \\
or & & and \\
\diagdown & & \diagup \\
 & term, \; truth \; value &
\end{array}
$$

Fig. 3.9 Naive logical relations (left) and corresponding mundane terms (right). Notation: S, E = proposition, + = or, * = and, ~ = not, f = false, t = true

propositional qualities:[20] S, E, ~S, ~E, f, t, where S = state, E = effect, ~ = not, f = false, t = true.[21] We use abbreviations: * = logical and, + = logical or, ~ = logical negation (cf. not). Terms separated by a comma denote synonymous meanings from some perspective. Just as the concepts of a state and effect are considered differently in each event, S and E indicate different logical meanings in subsequent events of the process model.

In *sorting*, the input collection is divided into two groups, representing qualities that are in focus (*state-in-input*, *effect-in-input*) and those that are complementary (*context*). We begin with an analysis of the first group. Because an effect always presupposes the existence of a state, *effect-in-input* has the meaning of co-occurrence and, therefore, a logical and-relation ($S*E$). To consider the input state as existence is identical to interpreting it as a constituent of the input interaction. Since, from a logical stance, state and effect are not different, *state-in-input* has the meaning of co-existence and, thus, a logical or-relation ($S + E$).

Which logical expressions can be assigned to the second group? They should be identical to those of the first group, except they inform about qualities that are *not* in focus. Just as the *context* represents background information as a single collection, complementary data about co-existence (logical or) and co-occurrence (logical and) are represented synonymously (~S + ~E, ~$S*$ ~ E). Readers familiar with propositional calculus may recognize these terms as logical not-and ('*nand*') and logical not-or ('*nor*'), known as the Shaffer Stroke and Peirce Arrow. The similarity of the relation between the two groups and between the two rules of De Morgan's laws must be clear. In Boolean logic: ~$(S + E) = $ ~$S*$ ~ E; ~$(S*E) = $ ~S + ~E.

In *abstraction*, the input state and effect are represented separately by *state* and *effect*. The first (*state*) means the input (S) as a possible subject for any suitable effect. The input effect (E) is excluded from that set since the relation between the input state and effect has yet to be established, so the input effect is not known as a quality suitable for interaction with the input state. By interpreting ~E, i.e., $not(E)$ as 'an effect not in focus,' the abstract state can be assigned the meaning of a logical

---

[20] As propositions and truth values.

[21] S, E, ~S, and ~E are unary relations (cf. logical variables); f and t are nullary relations (cf. logical constants or truth values). Unary and nullary relations have a single operand and no operand, respectively. S, E, and ~S, ~E logically represent information that is in focus and complementary.

inhibition relation $(S^* \sim E)$. Since state and effect as constituents are logically identical, the meaning of _state_ can be logically expressed in two synonymous ways $(S^* \sim E, \sim S^*E)$.

The second result of _abstraction_ (_effect_) informs, in addition to the effect, also about a state, as a consequence of the oft-cited property that an effect always presumes the existence of a state. Since the relation between the input state and input effect has not yet been established, and thus its meaning is unknown, the state involved by the effect must be different from and compatible with the input state, so it must be complementary $(\sim S)$. The co-occurrence relation of the effect (E) and the involved state $(\sim S)$ can be logically expressed by $\sim S^*E$. For reasons of treating state and effect logically identical, the above relation of the effect can also be given by $\sim E^*S$ or, due to the commutativity of the and-operator, by $S^* \sim E$. The two terms describe information by the effect (_effect_) from a passive $(S^* \sim E)$ and an active stance $(\sim S^*E)$. The combination of the two terms, $S^* \sim E + \sim S^*E$, known as a logical exclusive-or relation, expresses the alternative nature of the above two views, that the input can be viewed from the perspective of the state or the effect, but not both.[22] Note that a '+' symbol (logical or) indicates the logical meaning of a relationship in the sense of possibility.

The expressions by _complementation_ can be logically explained as follows. A state can exist in itself, and this also holds for the abstract state (_state_). Therefore, when a _state_ is completed with effect information in _complementation_, the state's logical meaning $(S^* \sim E, \sim S^*E)$ simplifies to $(S, \sim S)$. Also, completion of the abstract state (S) with an effect from the context $(\sim E)$ is in the sense of possibility. Accordingly, _state-in_-context can be logically expressed by $S + \sim E$ and, due to the logical identity of state and effect, synonymously so by $\sim S + E$. In short: $S + \sim E, \sim S + E$.

As for the _effect_, it is completed with data about the involved state in _complementation_. $S^* \sim E$ and $\sim S^*E$, which are dual following the logical identity of state and effect, simplify to $\sim E$ and E (see the analogy with the logical expression of _state_). In _complementation_, the latter (E) is completed with information about the involved state, compatible with the input state (S), and the complementary meaning of the first $(\sim E)$ with data about a complementary state $(\sim S)$. The possibility-sense combination of the expressions emerging from this event $(S^*E + \sim S^* \sim E)$ is the logical meaning of the abstract effect in context (_effect-in-context_) or the input effect in an active and a passive sense.

State and effect, as constituents of the input collection of qualities, are structurally related and, in this sense, imply one another. This dependence is expressed by $S^* \sim E$ and $\sim S^*E$ (_state-in-context_) as a logical implication relation and $S^*E + \sim S^* \sim E$ (_effect-in-context_) as a logical equivalence relation.[23] In math, an equivalence relation defines a set as a characteristic property. The link between _effect-in-context_ and _state-in-context_ as the predicate and subject of the input proposition must be

---

[22] Logical exclusive-or is true if only one of the operands is true; otherwise, it is false.

[23] In Boolean logic, $\sim S + E = S \rightarrow E$, $S + \sim E = S \leftarrow E$, and $S^*E + \sim S^* \sim E = S \equiv E = E \equiv S$.

clear.[24] The compatibility of the two representations is the meaning of the bi-directionality of implication ('$\rightarrow$' and '$\leftarrow$') and the symmetry of equivalence ('$\equiv$'). In Boolean logic: $\sim S + E = S \rightarrow E$, $S + \sim E = S \leftarrow E$, and $S \equiv E = E \equiv S$.

3.8.2   An Unexpected Conclusion

The final representation of the input interaction, by *predication*, explains why the input state (S) is affected by the input effect (E). The corresponding logical relation is a statement or proposition offered for reasoning: S *is* E. Remarkably, propositions functioning as a premise fall outside propositional logic. However, this is different from the unexpected conclusion announced in the title of this section. Before we get into that, let us summarize the naive logical relationships discussed, namely: S, E, $\sim S$, $\sim E$, t, f, $S + E$, $S*E$, $S* \sim E$, $\sim S*E$, $\sim S* \sim E$, $\sim S + \sim E$, $S* \sim E + \sim S*E$, $S + \sim E$, $\sim S + E$, $S*E + \sim S* \sim E$. Based on the process model, these relations can be arranged in a hierarchy (in technical terms, an induced partial ordering), which we call *naive logic*.

What is unexpected is the coincidence of the above naive logical expressions, in number and meaning, with the sixteen Boolean propositional relations on two variables, indicating the completeness of the process model from a logical perspective: We look at the input from *all* angles to determine an environmentally significant response to the stimulus.[25]

Let us note that a Boolean interpretation of the process model's events also illustrates the possibility of naive logic as a procedure. For example, by considering *complementation* using information from <u>context</u> as Boolean logical negation ('$\sim$'), <u>state-in-context</u> can be explained by $\sim(S* \sim E) = \sim S + E$, $\sim(\sim S*E) = S* \sim E$; and <u>effect-in-context</u> by $\sim(S* \sim E + \sim S*E) = S*E + \sim S* \sim E$.

3.8.3   The Brain Versus the Computer

The above result can be interesting, but why is it worth knowing? The point is that logical completeness and what it entails, the possibility of the process model being meaningful, can be practical for computer science. How it can, is far from obvious. There are two problems with computer use. First, the current von Neumann

---

[24]Note the naive logical meaning of the relation between *state-in-context* ($S + \sim E$, $\sim S + E$) and *effect-in-context* ($S*E + \sim S* \sim E$), expressing the latter as a naive logical and-relation (cf. co-occurrence) of the two terms from the first. In Boolean logic: $(S + \sim E)*(\sim S + E) = (S*E + \sim S* \sim E)$.

[25]It holds when all events of the process model are realized.

computer[26] is our only tool for efficient data processing. We have to use it, whether we like it or not. The second is that meaningful processing is qualitatively more complex than computation. Authentic interpretation cannot be achieved with the computer alone, at least not with the computer as we know it today. Without a theory of meaning, how can we justify whether the computer's result denotes an existing ('real') phenomenon or only makes sense formally?

The importance of naive logic is that it provides insight into the properties of meaningful processing. According to the hypothesis of this book, respecting the logical meanings and their dependencies can facilitate the experience of computer output as meaningful. Whether this conjecture holds and why remains to be seen. In any case, the analogy with apparent motion perception should be clear: Presenting the results of the information processing events in their induced order can allow the human agent to determine their collective meaning.

It is generally accepted that human interpretation is less efficient than information processing by the computer. But the emphasis is on something other than efficiency. More importantly, the brain interprets world phenomena as an interaction between a pair of independent qualities (cf., a relation between two variables). The computer suffers no such limitation. It can calculate relations on any number of variables but runs the risk that the result is only formally meaningful.[27]

3.9   Related Problems

Human-computer interfacing is not without problems. Formal mathematical explanations, also called derivations, can be long, tedious, and incomprehensible to the human agent. Experience with mathematical proof systems shows that formal derivations can easily use thousands of axioms and lemmas (cf. simple statements). Human communication also can suffer from long explanations. Speakers can use too many words to get their message across, and books can take a detour that contributes only marginally to the narrative and is seemingly included for artistic reasons only. Meaningful summarization may become necessary with increasingly lengthy reasoning, formally or by humans, due to the compelling need for efficient interpretation.

Besides the problem of voluminous data, processing efficiency imposes conditions on memory representation, which amounts to generalizing information into types. As an unexpected side effect, this allows, for free, a syllogistic interpretation of memory relations. These topics, namely summarization, memory representation, and a reasoning interpretation of stored data, are the subject of the coming sections.

---

[26] https://en.wikipedia.org/wiki/Von_Neumann_architecture

[27] And yet, the brain can also outperform the computer, not just because it is meaningful. According to a recent study, human brain cells can learn faster than some machine learning algorithms, surprisingly. https://futurism.com/brain-cells-play-pong/amp

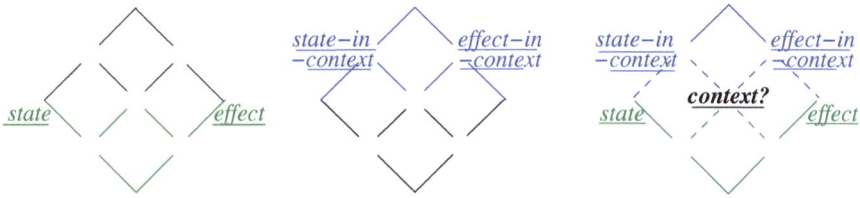

Fig. 3.10 Principle of summarization. Striped lines indicate relationships to be established; '*context?*' denotes complementary information used by the summary

3.9.1 Summarization

A famous statement, attributed to Albert Einstein and repeated by many, including Richard Feynman, says: "If you can't explain it to a 6-year-old, you don't understand it yourself." What is impressive about this quote is the belief that complex thoughts can be simplified. That this idea is by no means unfounded is shown by our ability to convert a series of utterances into a single statement. Think, for example, of a television program booklet presenting the content of a play, such as Romeo and Juliet, by a short text or even only a catchy sentence. Pushing text summarization this far has consequences, as the summary may not respect certain aspects, such as the play's narrative structure. Previously, we proposed summarization as an analog of apparent motion perception, where a series of static images is represented as a single quality. Based on this potential of the brain and the model of this book, we introduce a theory of meaningful summarization below.

Apparent motion perception can be considered a series of transformations from a previous image as a starting point to the next image as a goal.[28] The process of summarizing is analogous. Information from two process instances, previous and current, must be merged into one process, their summary. One way to achieve this is by establishing a relation between 'low-level' interpretation moments from the previous process (*state* and *effect*) and 'high-level' ones from the current process (*state-in-context* and *effect-in-context*). Because the two process instances refer to different phenomena, merging their information requires a context (*context*) enabling corresponding representations to be linked. See Fig. 3.10.

We skip details and illustrate summarization with our running example of mosquito bites. We assume that the first or 'previous' process provides *state* = *skin* and *effect* = injection. The resulting propositional expression, '*(there is) mosquito bites*,' is represented by a quale in the second or 'current' process. By assuming that the latter process defines *state-in-context* = *swelling* and *effect-in-context* = *malaria infection*, as well as a final expression of the input meaning, '*(a) malaria disease*,' we have all the ingredients necessary for merging. Context information enabling linking the two processes can be: *context* = *dangerous mosquito bites*. In this context, *skin* can be transformed into a *swelling (skin)*, and *injection* can be

---

[28] Whether apparent motion perception can be similarly explained is beyond our focus.

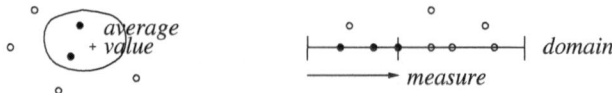

Fig. 3.11 Concise representation of data as a state (left) and a measure of an effect (right). Black/white bullets = responding/not-responding memory elements

interpreted as *(injection causing) malaria infection.* We ultimately get the proposition: *'The observed swelling of the skin may be a malaria infection caused by dangerous mosquito bites'*; in short, *'mosquito bites can cause malaria.'*

Summarization can have different results in keeping with the trial-and-error nature of finding a suitable context and a shared perspective for merging information from a pair of process instances. Not all summaries are useful. Many can be worthless. Some can be humorous, as illustrated by a dialogue from a Woody Allen movie.[29] In this scene, the actor says he has taken a thorough course in speed reading. "I read 'War and Peace' in 42 minutes. What's it about? I'll tell you what it's about. It's about Russia."

3.9.2 Memory Representation Revisited

The compelling need for efficient processing forces the interpretation process to generalize information into types. Since the set of memory responses activated by the stimulus can be considerable, time constraints applying to input activation can force elements of the set to be processed in chunks, not one by one. Chunks underlying the type concept are treated as a single value. What are the options for such a concise representation of stored information?

A set of state-type response data can be represented as a whole based on its elements as independent values. In Chap. 2, we called such a representation of the input state an *average value* (remember that the term 'average' is used as a metaphor for a mean or median of a set of properties, as in the concept of 'average man').[30] See Fig. 3.11.

Effect-type memory response information is not only a value but also has the meaning of a relation. The ground for the latter is again the property of effects to involve a state. Based on their qualia as values, effect-type memory responses referring to the same involved state can be arranged in an order, which we previously called a *domain.* Accordingly, the input effect can be compared with memory data from the same kinds of effect and, as a result, positioned as a *measure* of their domain. An example is knowledge about train noise of different intensities observed in the past (domain) and the input sound perceived now (measure). Because a

---

[29] https://quoteinvestigator.com/2015/12/08/speed-reading

[30] This value can only exist dynamically during information processing and can be learned and stored in memory. Figure 3.11 indicates this last option with a '+' symbol.

domain is a set of values, it can also take the meaning of an average value, like a state.

Since an effect can, in principle, represent any measure, effect domains must be continuous. However, computer memory is only suitable for storing discrete values, not continuous ones. For this reason, we are forced to model effect domains using *dense* sets. Elements of a dense set stand for effect measures observed in the past. New measures can be added by expanding the set by a new element and adjusting the existing order relation. In other words, observed measures are represented by an (existing) value, possible ones only as a potential value.

An example of a domain and existing values are 'locomotion-by-train' and 40 km/h (slow) and 180 km/h (rapid), respectively. An observed novel measure, 120 km/h (fast), can be remembered by inserting a new value between those existing ones. Values denoting a potential speed measure are not stored.

Because domains are an ordering, a measure can also be considered a *subset* of values, defined by all values less than and equal to the value indicated by the measure. Different values can have identical symbolic names as a measure. For example, 'fast' can refer to 120 km/h if it is about a train and 40 km/h if it is about a bicycle.

It should be clear that state- and effect-type information is interpreted differently. Does this mean that their information must be stored separately in the brain? That would undoubtedly be inefficient. We need evidence of how the brain gets around this problem. Still, the data representation assumed by our model may provide a solution. The essence of this approach is that although the two types of information are different, their representation can use the same memory organization, namely a connected network. A collection of memory data can be used to define an average value of a state; such a collection, this time as a linked structure, allows the definition of a dense domain and an order relation, facilitating the introduction of new values as measures.

Data representation by an average value and a measure of a domain are integral parts of all information processing events. For example, in *complementation*, the average value assigned to the input state is adjusted by complementary information used as a value. The domain associated with the input effect can be adjusted analogously as a set, in addition to updating its order relation due to the appearance of a new measure. The question of how the two types of data representation facilitate information processing is later on our agenda.

3.9.3   Reasoning

Average-value and measure representation are rooted in the conception of memory as a linked structure. A surprising feature of this structure is the ability to reason for free. The basic idea is simple. Because a link between memory information, as vertices, indicates an interaction observed in the past, the dependency between their data as state and effect can be seen as a relational expression of a proposition.

Reasoning can be understood as combining a pair of propositions, as premises, into a third proposition, as a conclusion.

As an illustration, we revisit our example of the sound of a fast-moving train in combination with the following conditions. We assume that (1) memory information 'train noise' and 'sound' respond to the stimulus in the sense of agreement[31] and 'loud' in the sense of possibility; (2) 'train noise' and 'sound' are not linked, and 'loud' is connected with 'sound' and 'train noise'; (3) the stimulus, first considered as 'sound' is eventually recognized as 'train noise' (the similarity of this condition with analysis mode processing is not accidental; more about this in the next chapter).

A syllogistic interpretation of the above memory relations is as follows. See Fig. 3.12. The first or *major* premise states information ('loud') about a memory element that responds in the sense of agreement ('train noise'). The second or *minor* premise postulates that the same information ('loud') also applies to another agreement-sense memory response ('sound'). The *conclusion*, joining the premises into a single proposition, explains the input activation as a combination of the aforementioned agreement-sense memory responses ('train noise' and 'sound') and shared information ('loud'). This result can be understood as a hypothesis: 'Train noise IS this sound,' or in mundane terms, 'It must be a train noise, this loud sound.'

An important conclusion we can draw is that memory organization can be the ground for the brain's reasoning ability. The term 'ground' has several meanings. We consistently use this term to indicate a phenomenon that can give rise to another phenomenon. An example is the rolling of train wheels on the rails (first phenomenon as ground) and train noise as a warning of danger (another phenomenon). No train noise exists without the interaction between the rolling wheels and the rails. Still, it is the train noise, not the wheels rolling, that a car driver may ultimately interpret as a warning of danger. Note that 'ground' is just a possibility; there is no plan or intention behind its existence. It is like Chekov's Principle: If a loaded rifle is on the stage, it will go off.[32] However, unlike a theatre performance, we cannot know if or when it will ever happen.

The importance of reasoning for human thinking is obvious. But it is also interesting for another reason. Experimental evidence shows that logical reasoning relies on conscious processing (DeWall et al., 2008). Combined with our analysis of memory as a structure underlying reasoning, this evidence implies that memory's architectural features can be the ground for consciousness. We should note that consciousness is a biological phenomenon, not to be confused with knowledge (Searle, 1993). Many states of consciousness, such as nervousness or undirected anxiety, have little to do with learning. It has yet to be discovered how consciousness can happen, but that it can happen needs no proof. Hypothetically, network memory

---

[31] The assumption of the existence of an agreement-sense response information pair follows from the condition that the input represented by interpretation is an interaction between (a couple of) independent qualities.

[32] https://en.wikipedia.org/wiki/Chekhov%27s_gun

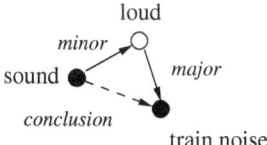

Fig. 3.12 Analysis mode processing as reasoning. *Major premise* = 'loud IS train noise' (loud is the usual intensity of a train noise); *minor premise*='sound IS loud' (this sound is loud); *conclusion* = 'this train noise IS sound (it must be a train noise, this loud sound). Black/white dot = high/low-intensity response; solid/dashed edge = existing/hypothetical relation. An arrow indicates the flow of information from the major to the minor term

organization can be the ground for the phenomenon of consciousness. It can be explained as follows.

If memory information responding to the external stimulus in the sense of agreement is satisfactory for a response, input processing can be efficient. The associated low energy consumption may explain why matching mode processing can be invisible to sensation and seemingly automatic. Interpretation may be less efficient when input recognition requires information that responds to the stimulus in the sense of possibility. After all, retrieving and combining complementary memory data with information in focus requires extra energy. This way of processing is characteristic of logical thinking, which involves integrating information from different domains because of its abstract general meaning. Speculatively, the rise in energy consumption, as a change, may function as a wake-up call for the brain, telling something significant has happened, and thus may explain why analysis mode processing reaches the realm of consciousness. It should be noted that the way of operation described is just one of the hallmarks of reasoning and that consciousness is more than a process.

The idea of two modes of operation by the brain, automatic and conscious, is distantly related to the 'dual process theory' invented by the philosopher and psychologist William James.[33] In his view, there are two kinds of thinking: associative and true reasoning.[34] The first is "only reproductive"; the second is useful for "unprecedented situations." The analogy with matching and analysis mode processing should be clear. The link between the two processing modes and reasoning is also trivial. In matching mode, the input phenomenon is recognized as a premise, which is a conclusion. In analysis mode, the input is defined by a pair of nested phenomena (cf. major and minor premise) processed as qualities. Modeling reasoning as a process is postponed until Chap. 5.

---

[33] William James was an American philosopher, historian, and psychologist, a close friend of Charles Peirce. https://en.wikipedia.org/wiki/William_James

[34] https://en.wikipedia.org/wiki/Dual_process_theory

3.9.4   Conceivable Consequences

It is tempting to think that architectural differences underlying matching and analysis mode processing can function as the ground for phenomena other than consciousness. How far can we push this idea? Let us note that the ground for a phenomenon can say very little about its meaning, so we must be careful.

In syllogistic reasoning, the major premise designates a general case, while the minor premise refers to a particular observation. The relationship between general information and stored knowledge allows for an unexpected conclusion. Suppose a statement appears in your consciousness as a thought, which you use as a major premise in a reasoning process. If this premise emerges from an interpretation process, it must be transformed into a quality; if it comes from memory, it can refer to data resulting from generalization. In any case, there is a chance of information loss. If, as a result, you cannot tell which individual phenomena the premise is about, you can conclude that it is based solely on your knowledge. You will experience it as a thought that is not forced on you and that, therefore, arises from *free will* (Churchland, 2006).

However, there is more to say about thoughts entering conscious experience unexpectedly. It is generally believed that conscious moments are interrupted by (automatic) non-conscious processes. As a result, at conscious moments, we are confronted with thoughts that we cannot link to the input, nor can we combine them with existing other thoughts through reasoning. The apparent logical discontinuity can be subject to experience and be the ground for the sensation of intuition. Because conscious moments are exceptional, most of our thoughts must be intuitive. In a broader sense, intuition and free will have a more complex meaning that lies beyond our horizon.

3.9.5   Interpretation Perspectives

We have seen that architectural properties are fundamental for the formation of knowledge. What else can be said about the properties of meaningful processing? Yarbus' experiment suggests that the change due to qualities shared by and different between successive images could be the basis for visual processing and, in a broader sense, for information processing in any domain. The collection of qualities can be extensive, and any subset of the input state and effect can be a source of change and interpretation from some point of view. How many different perspectives can be taken? Can we speak of an unlimited number of perspectives?

To answer these questions, we must delve into a theory of interpretation, particularly Peirce's framework of signs. Unfortunately, a simple exposition of his sign theory, semiotics, is unavailable. For the average man, Peircean theory is like a fortress whose main entrance is open only to the expert. Fortunately, there is a chink in another door, his theory of sign relations, which offers better perspectives. Through this gate, we can gain insight into the Peircean sign concept and his theory of categories, which is also the promise of the next chapter.

Chapter 4
Brain Processes Are Formed by Sign Processes

Abstract Starting from the idea that input stimuli function as a sign, we delve into an analysis of the properties of signification. Signs are more than just data: they are a transformational process. What are the events of this process? Because computations are limited to establishing a relationship on numbers, they are not transformational, at least not from the perspective of information as a number. If so, can the computer deal with signs? Can it create meaning? A theory of signification is due to C.S. Peirce. According to him, sign processing uses three types of relations that can be distinguished into three classes. Analytically, we show that the emerging nine sign relation classes encompass the meaning of the nine representation events introduced in the previous chapter. This implies that the process defined by those events can be meaningful.

4.1 What Does It Take for a Computer that Can Think Human-Like?

We can look for an answer in the physical properties of human processing, i.e., in neuro-scientific research. A drawback of this approach is that a theory of the brain that could be directly translated into an algorithm has yet to be found. Thinking is also related to meaningfulness. So, we can find an answer in philosophy. Unfortunately, the philosophical notion of meaningful processing is too abstract to apply in practice directly. Since communication with the computer is essential for our goals, but computer processing is limited to calculations on numbers, we face a challenging problem. How can we ensure that results by the computer are experienced as meaningful by the human agent?

Transforming the input data into an environmentally significant response, characterizing meaningful processing, is dynamic. When it comes to modeling thinking as a process, it poses a serious problem, as it is the same challenge we face with all natural phenomena. As Y.B., the unknown painter who tried to capture the essence of dynamic movement in a painting, once said, the difficulty is that it is always in motion. When faced with such phenomena, our usual reaction is to focus on properties and parts that we can recognize and, based on them, try to explain the

J. J. Sarbo, *Inevitable Knowledge*, SpringerBriefs in Computer Science,
https://doi.org/10.1007/978-3-031-73461-8_4

phenomenon as a whole. In the context of this book, it may be practical to concentrate on how meaning as a whole emerges from its parts as constituents, that is, how meaning is formed. Overall, it is a complex task, but if we succeed, we are done. By creating 'thoughts' based on the rules of meaning formation, the result should not differ from thoughts produced by the brain.

Unfortunately, they are qualitatively different. As the word suggests, formation is about form, not authentic meaning. Interpretation is a complex process that goes beyond a series of events. A series alone lacks the dynamics that fuse the events into a comprehensive input transformation into a response. It's like a cookie recipe, where ingredients (cf. qualities) are combined based on knowledge about their relational properties (e.g., that sugar can be mixed with milk) without knowing them in every sense (e.g., their chemical nature, how they taste). To follow the recipe means creating relations on qualities. Whether the end product is tasteful (meaningful), we cannot know, although there is a likelihood it is.

Experiencing form as meaningful is the core idea of our approach. If the rules of formation rules are derived from a theory of meaning, formation as a process could be experienced as meaningful by the interpreting agent. If this assumption is correct and, because we are such an agent ourselves, this approach to meaningful processing can work theoretically. But is there evidence that it can work in practice?

The success of the idea of form as meaning is known from role-play. By wearing the familiar white coat of the doctor, children can pretend to be such a person. Surprisingly, this may be enough to imagine that they are, which makes meaning as formation interesting for us. The good news is that even if we can only recognize the constituents and properties of a phenomenon, combining them allows us to experience their information as meaningful in most cases. There is also bad news. Although formation is less developed than meaningful processing, even that goal is not readily available to the computer. The metaphor of the cookie recipe perfectly illustrates the problem. The instructions tell what to do and in which order; in this sense, it is a formal process. Still, the recipe is only formally defined in some respects, as the ingredients are natural. Only when the ingredients are represented as numbers and operations on ingredients as calculations on numbers will the possibility of a computer that can think human-like come within reach.

In summary, to have a computer that can think human-like, we need a theory of meaning and, based on that, a model of meaning as a formation process and, finally, a computational representation of the events of that process.

We aim to show that the process model of the previous chapter is compatible with Peirce's sign theory (which we will explain shortly) and is suitable for computational implementation. The fact that the sign concept is based on a theory of meaning, but the process model is about meaning only as formation, the compatibility of the two views also illustrates the intimate relationship between content and form.

One final note. Since our focus is limited to the relation between the process model and Peirce's sign concept, the terms interpretation, meaningful processing, and sign processing can be used interchangeably.

4.2 Dynamic Interpretation

What does meaningful processing entail? Assuming that our thoughts arise from phenomena appearing in our perception, we can reformulate our question: How do phenomena convey information about the world? Broadening the perspective also serves another purpose: to convince the reader that thinking is not an 'invention' of the brain but a specific case of interpretation, the operation of the functioning sign. Since this latter question sounds more challenging, we will start with it and put our initial question aside, at least for a while.

Interpretation is about responding to what affects us. Since our thoughts appear to us in our experience, they must be phenomena. Hence, we have reason to begin with a theoretical treatment of phenomena to gain insight into interpretation properties. Such a framework can be found in Peirce's theory of categories. Because his category concept is too abstract to explain immediately, we defer its discussion to the next chapter. Before that, as a first step, we study Peirce's concept of interpretation, not from a philosophical but a process point of view. In line with that view, we offer an analysis of the events of that process and the types of distinctions they reveal.

Based on evidence from apparent motion perception, we assume that interpretation processes of all kinds are phenomena appearing in our experience. Regarding visual interpretation, we experience motion as a quality and the events of the underlying visual process that give rise to that quality. Motion perception is especially interesting for illustrating the brain's ability to convert static data into dynamic sensation by interpreting the difference between successive images as a change. Here, we advance the assumption that human processing is based on the same principle and can work analogously to apparent motion perception. The dynamics of interpretation can be captured in a process (see projection), and the properties revealed by the process in increasingly evolved representations of the input data (see images). However, the two phenomena differ in an important aspect. While the images come from an external source in motion detection, the interpreting system provides the input representations in meaningful processing.

The existence of a process alone does not imply that the emerging input representations are meaningful. To prove they can be, we consult Peirce's sign theory and ask how the response to the external stimulus arises through interpretation. The choice for his theory is motivated by the potential of the Peircean sign concept to generalize informational processes, including thought processes, into a single scheme. Since signs are phenomena, Peirce's category theory will come in handy to get hold of their meaning. To learn the properties of sign processing, we need to know how phenomena function as a sign and what information they convey. So, after this detour, we are back to our earlier question about interpretation, i.e., the nature of sign processing, which we finally get to work on.

4.3   The Naive Meaning of Interpretation as a Process

Let us start with how phenomena can function as a sign and fulfill their informational task. That meaningful processing is qualitatively more complex than a formal relation between stimulus and response is generally accepted. At the same time, it is also agreed that human processing should be simple. Anyone can do it with ease. What could be the secret of interpretation? Since we still do not have a clear picture of meaningful processing, we go back to our example of a mosquito bite and look at it from a broader context to see what kinds of meanings interpretation can reveal.

The example begins with a feeling of itching, accompanied by the visual signal of a red spot on the arm. In response, you quickly move your hand to scratch your elbow. Meanwhile, you accidentally hit your neighbor, who asks in that particular voice: What are you doing? You say apologetically that you suddenly had a feeling of itching. At first, you had no idea what was happening and where exactly, but later, you discovered a red spot on the soft side of your elbow and realized that an insect must have bitten you. It was not until you recognized the characteristic properties of the feeling of itching and its location that you concluded it must be a mosquito bite. Your first thought was to hit the mosquito, but you scratched your elbow instead because the red spot told you the mosquito must already have flown away. And, of course, apologies for accidentally hitting you.

That is what happened, so that must be the meaning of itching. Not exactly. On closer inspection, it is more precise to say that this is the meaning of itching in this particular case. This refinement is justified, as itching can have many other interpretations besides mosquito bites, such as tickling, eczema, and allergies. In summary, interpretation can be context-dependent. Also, the final response can result from a series of preliminary thought responses.

4.4   Peircean View

How can we achieve the meaning explained above, this time based on Peirce's sign theory? From a Peircean perspective, itching is a sign. The object of this sign, what it stands for and replaces, is biting by an insect. At least, this is what you first think of. Mosquito bites, your second guess, is a more developed idea of the object. The thought that a mosquito must have bitten you is the actual effect or the interpretant of itching functioning as a sign. As with the object, there can be other, more refined interpretants, such as a motor action of scratching the elbow. In summary, the interpreting system[1] can determine the object and interpretant by developing ever-better approximations of their meaning and, in this way, by interpreting the stimulus as a sign. Remember that to do all this, the interpreting system must first represent the stimulus as a potential for a sign or, simply, a potential sign. See Fig. 4.1. To test

---

[1] Peirce called it the Representamen.

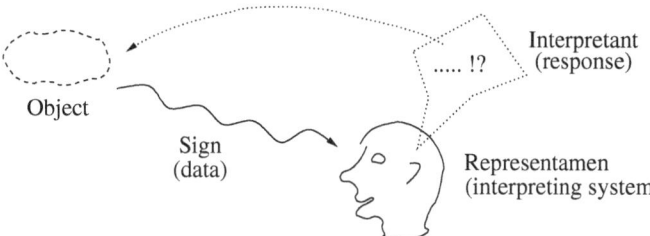

Object

Sign
(data)

!?

Interpretant
(response)

Representamen
(interpreting system

Fig. 4.1 A schematic picture of sign processing

whether this is clear, here is a question: What do you think is the meaning of the 'feeling' of itching?

If you think it is mosquito-biting, your answer is only complete in some respects. The correct answer is that the meaning of the 'feeling' of itching is the whole of its effects as a stimulus. Establishing this meaning requires determining three interrelated components: the stimulus, the object, and the interpretant. Omitting any of these determinations will change its meaning. Only the whole 'story,' i.e., the process of determination from stimulus to response, is what we can correctly call the meaning of the appearing itching. Later in this chapter, we will argue that this view applies to all signs.

However, there is a caveat with this explanation. The meaning of the stimulus is only evident if we know what is being affected. In the running example, we assumed that you, that is, your brain (including the sensory devices) recognizes the external stimulus. Your brain is affected by the stimulus (sign), which processes it into 'feeling' and finally into a thought or a motor action to scratch. Your brain realizes all these determinations, functioning as an interpreting system. The relief from scratching in the elbow is, at least in this example, beyond the effects of the current input. However, it may become a stimulus (and a 'feeling') in a subsequent interpretation process.

In summary, interpretation has three sites, i.e., where data is processed: the sign, the object, and the interpretant. Because their determination uses information stored by the interpreting system, sign processing demands the establishment of three increasingly complex relations: the relation of the potential sign or stimulus with (sensory) memory, the relation of the arising 'feeling' with memory about the object, and the relation of this 'feeling' with memory about the interpretant, in light of information about the object.

Peirce argued that the sign as interpreted, or simply, the sign, is defined by a relationship of the above three sites. Their determination, realized by the interpreting system, only exists within the process in which they function. In technical terms, they form an irreducible triadic relationship. The three determinations are experienced. The sign as 'feeling', the knowledge about the object, and the response (interpretant) arising from the sign's operations are our 'feeling,' knowledge, and response. By analyzing the triadic relationship of a sign into binary relations, the sign's meaning, hence the meaning of its three relations, is lost. Let us remark that not 'everything' is a sign. A potential sign will function as a sign only if it is interpreted as such, analogous to a process that only exists when it achieves its goal.

In light of the common idea that, as a result of interpretation, the stimulus can be transformed into a response based on stored information (knowledge), the Peircean sign concept does not sound strange. It can also be evident from a computational perspective, in which case a recipe or a program implements the determination of the three sites and memory information is defined as data. The three relations, as procedures, determine relations on the input qualities or, in the case of the computer, the program variables' values. All recipes and programs operate in this way.

It is not unexpected that a recipe-like process and, eventually, a computer program can be derived based on the Peircean view of interpretation. For a human-like thinking computer, the inverse problem is of interest whether the computer program (or process) is compatible with the Peircean sign concept. Not all processes meet these conditions. We will show that the process model of this book has this capacity.

A novel idea of Peirce's sign concept is a classification of sign relations and a theory of permitted ternary combinations of sign relations.[2] Equally important is Peirce's hypothesis that interpretation is not exclusive to the brain but is characteristic of all phenomena and, therefore, of nature.

4.4.1   Nine Relations

Let us return to how the stimulus can appear in our experience, i.e., what types of information it can convey. In the simplest case, the stimulus can occur as a 'feeling' or, according to categorical perception, a quale (*quality*). A more developed case is when we perceive the stimulus to occur with other qualities, for example, that the 'feeling' happens to us and appears now (*actual event*). Finally, we can perceive the stimulus as an effect we are familiar with, i.e., as an instance of a rule-like event (*rule*). *Quality*, *actual event*, and *rule* denote increasingly complex types of stimulus experience or, from a semiotic stance, the modes of operation of the interpreting system's relation with its memory and the types of information they reveal. The first (*quality*) deals with data about the stimulus in itself, the second (*actual event*), and the third (*rule*) with information about an ad hoc and a regular relationship of the stimulus with other qualities, respectively. No more types of stimulus experience are conceivable at this level of interpretation.

After recognizing the stimulus, the next task is determining what the stimulus (sign) stands for and replaces, that is, what its object is. The recognition of the sign's object also has different possible outcomes. Suppose the stimulus is considered only a *quality*. Since, in that case, we have no information about (co-)occurring other qualities nor about a rule-like meaning, only information about the stimulus itself,

---

[2]Ternary or triadic means 'consisting of three.' We use the two terms interchangeably.

such as its size, measure, or other formal (cf. form-related) feature of how the input looks like and is similar to can be retrieved from memory. As a result, the object can only be determined in this sense of *likeness*. When the stimulus is experienced as an *actual event*, referential information of the stimulus, in combination with its meaning as *quality*, allows the interpreting system to access memory about qualities that can co-occur with the stimulus, i.e., qualities that the stimulus refers to (*connection*). In addition, recognizing the stimulus as an *actual event* allows the interpreting system to determine the object only in the sense of *likeness*, thus irrespective of its meaning as a *connection*. An example of the latter is the 'feeling' of itching (*actual event*) as a measure on some scale (*likeness*); an illustration of the first is a pain in the elbow (*connection*). The third case of object determination is when the stimulus, recognized as a *rule*-like effect, allows the interpreting system to retrieve memory data about a *conventional* property that characterizes the object. In addition, analogously to the previous case, the interpreting system can determine the object in the sense of *connection* and *likeness*, irrespective of its meaning as *convention*.

Which combinations of the three kinds of information about the sign and the object are permitted? The interpretation process can know as much about the object as the stimulus information allows. So, the combination of *likeness* and *quality* is permitted, as opposed to the combination of *connection* and *quality*, due to the latter's lack of a referential meaning.

The last task in interpretation is to determine the response based on data from the stimulus and knowledge about the object. If the stimulus is recognized as *quality* and the object as *likeness*, then memory allows the response to be determined only as a qualitatively possible (*qualitative possibility*). As for the remaining cases of stimulus recognition, we only mention the most developed combinations of sign relations. If the input is recognized as an *actual event* and the object is determined in the sense of *connection*, then based on the memory data, the response can be characterized as an actual change in an existent (cf. subject), which is affected by the stimulus (*actual existence*). Finally, suppose the stimulus is recognized as a *rule* and the object in the sense of *convention*. Since, in that case, we are familiar with the input stimulus and the conventional meaning of what it stands for, the response must be clear. It can be determined as a statement or a proposition (*argument*).

In summary, the relations of the sign have nine classes or modes of operation, called *nine relations*. Considering that the stimulus can retrieve from memory only as much information about the object as the stimulus as a sign enables, and about the response, as much data as the object allows, it follows that from the $3*3*3 = 27$ ternary combinations, only ten are permitted.[3]

The set of dependencies of the three operational modes of a sign relation defines an induced ordering. It is illustrated by the relation of the interpreting system with its memory: an *actual event* presupposes the existence of a *quality* and a *rule* the occurrence of an *actual event* as its instance. The ten permitted combinations of sign relations also define an ordering. The two orderings can be merged. From a

---

[3] Ternary or triadic means consisting of three. We use the two terms interchangeably.

Fig. 4.2 The transitive reduction, a.k.a. the Hasse diagram (A Hasse diagram represents a finite partially ordered set in the form of a drawing in which all edges corresponding to a transitive relation (cf. a 'shortcut') are omitted.) of the set of dependencies of the nine relations. Right diagonal edges denote the ordering induced by the three sign relations. Left diagonal and horizontal edges indicate the orderings induced by the ten permitted combinations of the nine relations

mathematical stance, the set of dependencies suffers from redundancy due to dependencies representing a 'shortcut,' formally called a closure. When the nine relations are only used as types of information content (data), such as in the process model, closures provide no additional information and, therefore, can be omitted. See Fig. 4.2. The reduced set of dependencies plays a central role in proving the meaningfulness of the process model.

The significance of Peirce's sign concept for AI and IA lies in the definition of sign relations and their classification into operational modes. Those modes and their dependencies can be learned as data types and relations (see combinatory properties). Based on their information, memory knowledge can be organized.

Remember that the three sign relations exist only within the triadic relationship in which they function. A similar restriction applies to neural activity triggered by the stimulus in the case of the brain as an interpreting system. When the trigger ceases, the associated brain activity stops sooner or later. Outside of the ternary relationship, a sign relation is limited to a type of information content, such as a formal relation on qualities.

In comparing the Peircean sign concept with the process model, we treat sign relations in the above, restricted sense, and representations by the process as types of information content. A successful comparison implies that the process model can be embedded in the Peircean sign concept and, in this sense, can be positioned as meaningful. To that end, we will show that representations by the process model can be embedded in the nine relations. It implies that everything that the process model is capable of can also be achieved through signs. Since the sign concept is inextricably linked to interpretation, we finally found the key to the computer that can think human-like.

Well, not really. Embedding is an asymmetric relation.[4] The fact that the process model can be embedded in the sign concept does not mean the opposite is true. Knowing that signs are qualitatively more complex than a series of events, we may

---

[4] In math, embedding is defined by an injective, structure-preserving mapping.

wonder why the above embedding relation would be attractive for our goals. There are two reasons for this. First, the process model is a plausible description of sign processing in terms of its basic properties. While this still does not guarantee that it can be meaningful, evidence in modeling language processing and reasoning suggests it does. Second, due to its relational nature, the model suits a computational implementation, meaning that a solution to a human-like thinking computer can be within reach.

4.4.2 An Example

Before we go any further, we would like to see how the concepts introduced so far apply to our running example of mosquito bites. Information processing begins with perceiving the external stimulus as quality, i.e., as a 'feeling' of itching and also as a 'feeling' somewhere (in the elbow), that is, related to something else. By transforming the input 'feeling' into a response, interpretation determines increasingly developed meanings of the object, that is, what the stimulus stands for and replaces. The first or immediate object is an insect bite. A dynamically obtained, more developed, or simply dynamic object is a mosquito bite. The response to the stimulus, which in a sense responds to its effects, that is, the interpretant, can be a thought, for example, itching from mosquito bites, or a motor action, e.g., scratching the spot where the insect has bitten.

What can be said about the interpretation process? The 'feeling' of itching and the object of this quality were there before they were recognized in a response. Only through that response that relates the 'feeling' to a mosquito bite is the stimulus hypothetically recognized as incipient itching. In other words, the 'feeling' is transformed into a relationship with the object and response. This object, recognition process, and response define the stimulus as a sign. Information processing may not stop here. The response, and therefore the emerging ternary sign relation, can become a quality, hence a potential sign in a subsequent interpretation process.

4.5 Relation Embedding

Let us examine the dependence between the sign relations of the semiosis process and the process model's interpretation moments. See Fig. 4.3. In comparing them, the nine relations function within a ternary sign relationship, and the nine interpretation moments are types of information content. Here is an example. The sign relation *quality* denotes the interpreting system's relation with its memory in which the (potential) sign is considered data. Only in combination with the sign relations determining the object and interpretant can *quality* establish a ternary sign relationship. The question of how those two other sign relations are determined is irrelevant to our goals. In our analysis, we will assume that they exist.

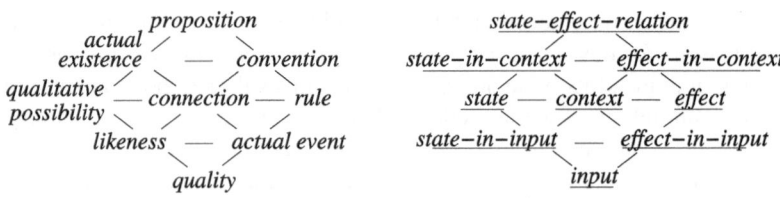

Fig. 4.3 The Hasse diagram of the set of dependencies of the nine relations ordering (left) and the relational structure defined by the interpretation moments of the process model (right) (The similarity of the two structures speaks for itself. To prove that the process model's dependencies can be embedded in the nine relations, as an ordering, we need to represent the process events as binary relations (in addition to embedding the process representations into the nine relations, as described in this chapter). These binary relations are depicted by horizontal edges in the right-hand side diagram in this figure.)

The input representation corresponding to *quality* is *input*. The dependence of the two concepts can be illustrated with the processing of an image (potential sign) in apparent motion perception. The sign relation *quality* denotes linking the input image with (sensory) memory information. The input is treated as data, and the arising memory response is data. In the *input* event, providing *input*, the interaction between the input visual signal and the interpreting system (e.g., the retina) is represented as quality, such as an ingredient in a recipe. It is natural, but in the recipe, it is treated only as relational information (what its properties are and which ingredients it can combine with). The fact that relational information from *input* can be embedded in the stimulus' meaning as *quality* must be clear (we assume that in the perception of a series of images, a single image is treated as a quality).

A comparison of the sign relations and interpretation moments can be tedious due to the number of cases, but the result is simple. It says information from the nine interpretation moments can be embedded in the nine relations. Proving this dependence is the subject of the next three sections. Readers only interested in the main ideas can skip them and continue reading afterward from Sect. 4.4.

The sign itself determines the three types of relations. In human processing, those relations are the result of the brain. So, the sign and the brain functioning as an interpreting system are the same. Based on this observation, the potential sign's relation with sensory memory amounts to a relation of the interpreting system with itself (woken up by the sign/stimulus, the interpreting system determines what happened to it in a 'feeling'). The relation of the sign with the object and the interpretant can be dealt with analogously. In short, since we focus on the interpreting system, we will consider the three relations of the sign as the relations of the interpreting system with itself, the object, and the interpretant.

4.5.1 The Relation of the Interpreting System with Itself

The first mode, *quality*, is when the stimulus (potential sign) is considered only data. We have just explained how it can be compared with *input* so we can move on to the next mode. The second mode, *actual event*, is when the stimulus is considered as a reference, in addition to its meaning as data. Can data function as a reference? For the computer, this may not be a problem; it can use almost any data as a memory address. For the process model, the question is whether the input data can refer to something else. That it does, is apparent from the *sorting* event, in which the input is represented as an effect (*effect-in-input*). A property, which we have mentioned several times, is that an effect always presupposes the existence of a state. Because of this relationship, all effects have a referential meaning. Since *sorting* represents the state only as a constituent, which condition also applies to the state involved in the effect, the referential meaning of the effect cannot exceed the possibilities of a pointer. How can an *actual event* be seen as a pointer? The input interaction appears in our experience and thus occurs now. Since the interaction is between an effect and a state (i.e., the stimulus and the interpreting system that occurs in some state), the collection of qualities defining the input must represent a co-occurrence event. By affecting the state, the effect must refer to and, therefore, point at the state. An event, now taking place, and a pointer reference are both involved in the meaning of an actual event phenomenon. This way, *effect-in-input* can be embedded in the sign relation *actual event*.

The third mode, *rule*, is when the stimulus is considered a habitually known effect. Again, we ask how the input data can have this meaning. What makes this relation possible is the event of *abstraction*, representing the input effect in a general sense (*effect*), independent of the input state. The arising abstract meaning is limited to learned properties of how the stimulus can affect any state (active sense) and how a state can be affected by the stimulus (passive sense). The two kinds of meanings define the input effect in a recipe-like sense. In this way, *effect* can be embedded in the sign relation *rule*.

4.5.2 The Relation of the Interpreting System with the Object

The first mode, *likeness*, is when the object is determined based on the stimulus as data. Regarding the process model, how can the object be identified based on data alone? Again, this is enabled by *sorting*, in which the input state is represented as a constituent (*state-in-input*). This information is based on a relation of similarity, just as a part is similar to the whole of which it is a part. As a constituent, the state indicates what is affected. It informs about the object of the input effect in this sense. That a relation of similarity is involved in the meaning of likeness goes without explanation. In this way, *state-in-input* can be embedded in the sign relation *likeness*.

The second mode, *connection*, is when the object is determined based on the input data as a reference. How can the object be identified by data used as a reference? The *sorting* event also makes this input ability possible, representing the input from a complementary stance (*context*). Knowledge about past interactions similar to the current one can be selected by the input data used as a pointer. Although the response information cannot explain the reason for the input interaction, being similar to the input, it can tell what is affected and what that effect can be. It can identify the object of the stimulus referentially by pointing to its stored data. That a pointer-like referencing is involved in the meaning of a connection should be clear. In this way, *context* can be embedded in the sign relation *connection*.

The third mode, *convention*, is when the object is determined based on the input data, seen as rule-like information. An expression of the object in this way is facilitated by the event of *complementation*, representing the effect in context as a characteristic property of the input state (*effect-in-context*). Since complementary information arises through learning and thus generalization, it can only identify the sense in which the object is affected, how it is commonly known, that is, as a conventional property. An example is a red spot on the skin. We can call it 'mosquito bites' because we commonly agree to designate such red marks by this symbolic, conventional name. In this way, *effect-in-context* can be embedded in the sign relation *convention*.

4.5.3   The Relation of the Interpreting System with the Interpretant

The first mode, *qualitative possibility*, is when the interpretant is determined based on the stimulus considered as data. How can the interpretant be identified based on data alone? Such a meaning is made possible by *abstraction*, in which the input is represented as an abstract state (*state*). The abstract state, independent of the input effect, can determine the stimulus's effect on the input state only in the sense of possibility, not as a possible for all effects, but only for those that can interact with the state similar to the input one, thus recognized in the sense of *likeness*. Based on information about such effects and the input state (object), a response to the stimulus can be determined in a recipe-like sense. A possible response is involved in the meaning of the state as qualitatively possible (for an effect). In this way, *state* can be embedded in the sign relation *qualitative possibility*.

The second mode, *actual existence*, is when the interpretant is determined based on the input as a reference. Again, this is enabled by *complementation*, representing the abstract input state in combination with data about its known properties (*state-in-context*). The resulting expression describes the input state through its actual properties, i.e., as an existing and thus actual state, commonly called the subject. In this way, *state-in-context* can be embedded in the sign relation *actual existence*.

The third mode, *proposition*, is when the interpretant is determined based on the input seen as rule-like information. Such a representation of the input is enabled by *predication*, which, by combining the input state and effect in context as subject and predicate, expresses the input meaning by a statement (*state-effect-relation*). A statement, also as an act, is involved in the meaning of a proposition. In this way, *state-effect-relation* can be embedded in the sign relation *proposition*.

4.6 Pending Questions

A sign is a relationship of three sign relations, each of which can be distinguished into three classes. Although this is clear, we still have a few questions. Why are there always three of everything? Why is there not a fourth, fifth, etc.? And why should the three modes of operation of a sign relation be qualitatively different? It may come as a surprise, but the two questions are not unrelated. However, we must cover more of Peirce's theory to address them adequately.

As a first step toward this goal, we begin with a problem we can immediately solve. It concerns an essential difference between Peirce's sign concept and the process model. Unlike the process model, which requires all nine interpretation moments to be determined (in analysis mode processing, at least), each of the three relations that make up a sign operates in one of three possible modes. This apparent contradiction can be explained easily through the naive logical interpretation of the process model. The goals governing sign interpretation (semiosis)[5] and the information processing model of this book (process model) are different. The first aims at establishing a ternary relationship of sign relations, and the second strives to describe the input interaction from all possible perspectives (see 'naive' logic). The process model can represent sign relations as types of information content; it cannot express them as a ternary relationship.[6]

---

[5]Peirce described semiosis as any process that involves signs. https://en.wikipedia.org/wiki/Semiosis

[6]For obvious reasons, Peirce does not answer the question of which of the three possible realizations of a sign relation is created in the case of a given sign and by what mechanism. The process model does not answer this question either. It does, however, reveal a possible mechanism at work.

4.7   How Can We Get a Computer that Thinks Human-Like?

How far are we from the ideal of a computer that can think human-like? And what other characteristics, besides being a process, can be attributed to the Peircean sign concept? Let us begin with the second question, particularly the conditions for meaningful processing.

Based on the similarity between the process model and the Hasse diagram of the nine relations as information content, events of the first can be seen as information increments according to dependencies from the second. Simply put, the process 'knows' more after an event than before. In technical terms, the sign relation assigned to the result of an event is higher in the ordering than the sign relations corresponding to the constituents of the event. Since input expressions obtained by the process are static information, but sign relations have a dynamic meaning, the condition for an event to be seen as a meaning increment is the ability of the interpreting system or agent to experience the process events as dynamic transformations. This ability must be available for two reasons. The first is the similarity mentioned above, as well as the relation between representations by the process model and the modes of operation of the semiosis process. The second, which is hypothetical, is evidence from apparent motion perception. Given the brain's ability to perceive static images as motion, it is not unlikely that input representations by the process model can analogously be experienced as a (meaningful) sign. By now it should be clear what may justify this analogy. The events of the process model can be viewed as state-effect relations. An example is the event of *complementation* of state by context (see Fig. 3.6), where data from the latter supplements, thus affects, data from the former. This relation, which represents the input interaction, must be an interaction so that no input information is lost. The change brought about by a state-effect interaction can be determined by comparison, just as in apparent motion perception. Recall that the condition for interaction is the occurrence of qualities that are independent but also have something in common. Accordingly, state and effect can be defined in terms of qualities that are common and those that are different. Research by Robert Goldstone and Lawrence Barsalou (1998) also emphasizes the importance of comparison operations by the brain. However, the idea that it uses an identical mechanism for information processing in a broader sense is hypothetical.

Then, there is a crucial question of which combinations of the nine relations define a ternary sign relationship and what makes that possible. Peirce's sign theory defines the permitted combinations of sign relations but does not explain how the modes are determined. If so, who or what is responsible for deciding which three of the nine relations of a sign will be merged into a ternary relationship? We assume that this task is left to the interpreting system. It is tempting to think that the ability to select at a higher level of information processing is related to the brain's capacity for selective attention, a cognitive phenomenon we mentioned earlier.

In summary, the computer that can think human-like must meet three conditions. The first is that its program, as a process, can be embedded in the nine relations dependency. The second is that information increments by the process can be understood as meaning increments. The third is that the process can make selections, facilitating input expressions to be combined into a ternary relationship. A computer implementation of the process model can only satisfy the first condition; the second and third conditions are beyond its capabilities. Fulfilling them can be entrusted to an interpreting agent. The process model may facilitate human processing and increase human intelligence in this way, hypothetically.

4.7.1   Complex Phenomena

Admittedly, the examples introduced so far are too simple. What is required for processing complex phenomena? What this entails, is illustrated below using the earlier analysis of mosquito bites, this time in a broader context. In short, the 'feeling' of itching, as a phenomenon, is embedded in the phenomenon of the visual signal of a red spot from mosquito bites. Later, we will argue that processing complex phenomena in natural language syntax, such as a closed clause, can be modeled analogously, i.e., via nested processing.

Embedding, transforming a response, e.g., a thought, into a single quality, can come with a loss of information; this is the price of processing only one phenomenon at a time. But that is not our concern at the moment. We welcome the idea that simple and complex phenomena can be processed identically, which implies that human processing can be based on *one* type of process. Also, it makes the difference between the brain, which, conceptually, is based on one procedure, and the computer, which can execute any program, even more apparent.

By now, we have an image of interpretation and the types of information its sign relations can reveal. However, we have a few open questions. Why are these relations? Why is their number limited to 3 and their dependencies to 3*3? Since interpretation deals with phenomena, we need to know more about the kinds of phenomena to answer these questions, which is the topic of the following chapter.

Chapter 5
Sign Processes Are Formed by Categories

Abstract Signs are a dynamic process transforming the input stimulus into a response, but what is the basis for this assumption? Why are there three types of sign relations, and why can they be classified into three classes? We find answers to these questions in Peirce's category theory, his classification of phenomena. In addition, we show that Peirce's categories explain the possibility of unified information processing. Proving that categories and the nine sign relation classes are practical, we discuss evidence of their use in various areas of human knowledge, including language processing, professional integrative negotiation, and mathematical problem-solving by elementary school children.

5.1 Classifying Phenomena

So far, we have taken it for granted that signs are dynamic processes, but what is the basis for this assumption? An answer can be found in Peirce's category theory, his classification of phenomena. Peirce's categories can be further used to explain the hierarchical arrangement of the nine relations and the possibility of uniform information processing.

A category denotes a class of things with specific common characteristics, but the related philosophical concept is more general. The word phenomenon comes from the Greek *phainomenon* and means "a thing that appears." Phenomena, however, are more than a thing. They appear in our perception via interaction, so they must be dynamic. What about static phenomena? Can 'things' exist? Of course, they can. Phenomena that do not change are what we can rightly call a 'thing.'

Perceived phenomena can be classified according to their properties, which is no surprise. However, the fact that the number of categories can be limited to three can be more unexpected, although the attentive reader might have already guessed from the many hints that there should be three of 'everything.' However, what is entirely new is the nature of the Peircean categories.

Interest in categories as the highest kinds or genera dates back over two thousand years. More on this soon. For now, it is sufficient to know that based on their properties, 'things' are traditionally distinguished into fundamental and different

J. J. Sarbo, *Inevitable Knowledge*, SpringerBriefs in Computer Science,
https://doi.org/10.1007/978-3-031-73461-8_5

classes. However, a classification that can handle the dynamic nature of phenomena requires a different approach. To this end, we revisit the Peircean sign concept and examine its dynamic features.

We already know that the ternary relationship of the sign is based on its three relations that transform the potential sign as data into a response based on knowledge about the object. Also, we have encountered the three sign relations in different contexts so far. An example is perceiving the signal of train noise by an arriving train as input data, recognizing the input data as known information, and providing a response. Another example is syllogistic reasoning, in which the minor premise (input data) is converted into a conclusion (response) based on knowledge from the major premise (object). We note that a similar distinction characterizes the precursor of the process model, the three-stage interpretation model. In stage (1), the input is distinguished into state and effect qualities; in (2), the two types of qualities are linked with stored knowledge; and, in stage (3), their meaning in context is merged into a response.

5.1.1   Three Questions

Why these relations of the sign? Why these stages of operation? And what is shared by the two concepts? We begin with the second question concerning the stages of operation. If, as we think, there is no homunculus, then the idea that those operations are forced upon the brain must be wrong. A more likely explanation is that they are what the brain does as an information processing system. Yarbus' experiment proves that interactions are essential for human processing. Based on this evidence and the fundamental nature of visual processing, we assume that interpretation is inherently related to interaction and, to avoid information loss, this must also apply to the result of that process. We already know that all interactions presuppose independent qualities, and both input and response can be characterized as phenomena. From these conditions, and because the response can function as quality in a subsequent interaction, we conclude that all phenomena must be an interaction and, thus, a duality. See Fig. 5.1. In this diagram, 'noise' and 'senses' are a duality, and their interaction, 'sensory signal,' must also involve independent qualities and thus be a duality.

The assumption that phenomena can be characterized through independent qualities holds in a broader sense: mechanical phenomena by distribution and intensity of energy; chemical phenomena by dilution and solubility of the chemical bond; wave

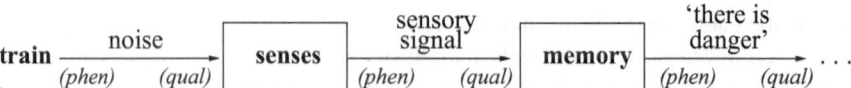

Fig. 5.1 Sample sign phenomena. Arrows indicate an interaction, e.g., between 'noise' and 'senses.' Signs appearing as a phenomenon (*phen*) can later function as quality (*qual*)

phenomena by frequency and intensity of a continuous stream; and information phenomena by form (cf. code) and content of a carrier signal.[1]

We cannot prove that the process model (and its three-stage variant) is fundamental to human interpretation. What is within our reach is illustrating their potential for modeling information processing in different domains of knowledge, including natural language syntax, syllogistic reasoning, and mathematics (numbers), and that is precisely our plan.

Regarding the first question at the beginning of this section about the nature of sign relations, we follow Peirce, who maintained that the basis for the three sign relations is the modes of organization of phenomena. It is those modes that he called a *category*. Also, concerning the last question, whether the two theories of interpretation share a common element, we can find an answer in the categories, suggesting that human processing and knowledge must be organized around the Peircean categories.

5.2   Categories

An attempt to define general classes for entities[2] goes back to Democritus's analysis of statements, from five centuries B.C. Aristotle introduced a systematic category theory almost two centuries later based on the same idea. For Aristotle, a statement is about something in the world, and if the world is as it is described, then the statement is true; otherwise, it is false. This view is known as the 'correspondence theory of truth.'

To show how a statement (an assertive sentence) corresponds to what is the case in the world, one has to specify which parts of the assertion refer to which elements in the world. Establishing this relation requires an analysis that shows the constituents of sentences and a study that reveals what can be distinguished in the world. Regarding the second, Aristotle assumed that there are entities on the one hand and properties on the other (Seuren, 1998). According to him, many kinds of entities exist, including properties that can be considered entities.

Properties can be ordered into ten basic classes or categories: substance, quantity, quality, relation, place, time, position, state, action, and affection.[3] Regarding the first, statements are analyzed in terms of a subject and a predicate. For Aristotle, the

---

[1] Dilution and solubility refer to the independent qualities of the diluted substance on the one hand and the liquid in which the substance is dissolved on the other. Not all pairs of qualities can be interpreted as a duality. For example, the two types of energy distribution, potential and kinetic, are not independent and, therefore, not dual. This paragraph is based on research by J.I. Farkas. See Sarbo et al. (2011).

[2] Merriam-Webster defines 'entity' as independent, separate, or self-contained existence.

[3] In more detail, the ten categories are substance, quantity, quality, relation (relatives), place (somewhere), time (sometime), (being in a) position, state (being acted upon), action (acting), and affection (having). https://plato.stanford.edu/entries/aruaristotle-categories/

Fig. 5.2 Kant's table of
categories

Quality	Quantity	Relation	Modality
affirmative	universal	categorical	problematical
negative	particular	hypothetical	assertoric
infinite	singular	disjunctive	apodictic

subject refers to an entity, while the predicate denotes a property. Analogously to
properties, predicates can also be classified under ten general categories. A statement
is a linguistic structure in which the speaker assigns a property to an entity (or set of
entities). The assertion is taken to be true just in case the property denoted by the
predicate adheres to the entity (or entities) referred to; otherwise, the statement is
false.

Truth correspondence theory is an expression of the fundamental nature of
experience. This view is reflected by the original meaning of the term category,
designating that which can be said, predicated, or publicly declared and asserted
about something, hence relying on common experience shared by all. Experience is
fundamental, but it has limitations. A lack of critical analysis has consequences. The
Dutch comedian Herman Finkers puts it this way: "The earth is round ... like a
pancake."

In Aristotelian theory, the entity carries the property assigned to it; the relationship between the subject and the predicate is static. Although Aristotle sets out the
classes of this relation, he does not question how they arise. Later philosophers, of
whom we only mention Kant, all adhere to the Aristotelian view of the world. Peirce
was the first to give a paradigmatically new answer.

Kant[4] investigated the different kinds of judgments (cf. statements of belief or
assertion) with which we think about objects. He introduced a table of four dimen-
sions: Quantity, Quality, Relation, and Modality, with three subdivisions in each
dimension. The idea is that, if classified, each judgment will score in each dimension
on one of the sub-divisions. To apply to objects, Kant generalized the table of
judgments, in a table of categories of being. See Fig. 5.2.

His theory is relevant to our purposes only because its analysis leads us to the
Peircean category concept. Peirce recognized early in life that we cannot start with
the notion of judgment in our quest for the most general categories.[5] Kant's table did
not convince him since he found too many faults in, and interrelations between, the
different categorial sub-divisions. He concluded that there must be a more general
level at which we have to search for categories. That led him to search for a
generalization of Kant's sub-divisions in triads. Peirce assumed that if we are
looking for genuinely universal conceptions, we have to find out what is needed to

---

[4]Kant believed that the ability of human understanding (Latin: ratio) to think about and know an
object is the same as making a spoken or written judgment about an object. Kant created a table of
the forms of such judgments as they relate to all objects in general and used that as a model for the
table of his twelvefold (4*3) categories. https://en.wikipedia.org/wiki/Category_(Kant). The current
and following paragraph is based on research by A.J.J. van Breemen. See Sarbo et al. (2011).

[5]By rejecting the proposition to be fundamental, Peirce refutes Descartes's famous claim, "I think;
therefore, I am," and states the opposite as true.

bring the manifold of sensuous impressions or the content of consciousness to unity, the unity of the judgment. It is in the formation of the judgment, not just the result, the judgment itself, where we have to look. The morphological structure of information, 'in' + 'formation,' i.e., bringing in form, also reflects this view.

5.2.1 The Dynamic Nature of Phenomena

Heraclitus (fl. 500 BC) maintained that struggle and change are the natural conditions of the world. He is known for his famous claim of *"Panta Rhei,"* ancient Greek for 'everything flows.' In a more profound statement, of which usually only the first half is quoted, he says: "No man ever steps in the same river twice, for it's not the same river, and he's not the same man." To come to grips with Peirce's category concept, it is worth analyzing what Heraclitus claims about the nature of dynamic phenomena. On a closer look, Heraclitus refers to three kinds of experiences. The first is the experience of water flow in itself. The second is the experience of 'otherness,' i.e., that water flow differs from us. The third is the experience of water flow as a familiar rule-like property. Why rule-like? Even though the flow is constantly changing, and we too change with time, if only because every observation can increase our knowledge, we can recognize the river as it is now, based on our general understanding of what water flow is. In each of the three experiences, water flow has a different meaning. In the first, it is a quality; in the second, it is an actual property; in the third, it is a habitual property. The three experiences are not independent: the experience of water flow as a quality is assumed by its experience as an actual property, which, in turn, by its experience as a habitual property.

The simplest case of a transformational process is a brute reaction, illustrated by Newton's cradle. See Fig. 5.3. The moving metal sphere (left) clashes against the stationary spheres (right). In the interaction, kinetic energy is transformed, via arising force and counterforce, into a movement by the second, third, and finally, the last sphere. The spheres 'experience' the appearing external power as an effect and transform it into a reaction, their 'response.' Their ability to react can be understood as 'knowledge' of Newton's third law of motion. It is important to note that the rule of "for every action in nature, there is an equal and opposite reaction" limits the meaning of reaction (response) to a relation. The true meaning of Newton's cradle is the emergent movement, the reaction of the spheres to the force due to the external kinetic energy.

Another illustration of brute reaction is a wave phenomenon arising from an interaction between waves. According to the Huygens-Fresnel principle, every point on the wavefront is a source of wavelets. Those wavelets spread out in the forward direction at the same speed as the source wave, transforming the source wavelet into a new wavefront line tangent to all the wavelets. As with Newton's cradle, the true meaning of a wave phenomenon is not the Huygens-Fresnel principle as a relation; it is the emerging wavefront, the response to the input wave interaction.

Fig. 5.3 Newton's cradle

Our running example of mosquito bites demonstrates that a meaningful response is more than a brute reaction. The visual signal of the red spot on the skin is converted into scratching the arm based on knowledge of similar observations in the past. The entire transformation, including the reaction, is the meaning of the mosquito bite as a sign.

5.3 Peircean Categories

Aristotle was aware of the dynamic nature of phenomena. In Metaphysics (Aristotle, 1984) he writes: "The totality is not, as it were, a mere heap, but the whole is something besides the parts [. . .]", which is often quoted as "The whole is more than the sum of the parts." In modern science, this is known as the concept of emergence. An emergent property of a system is not a property of any component but a feature of the system as a whole.[6] How exactly emergence works remains to be explained.

Peirce argued that we come to know the world through phenomena (Rosensohn, 1974), i.e., "whatever is present at any time to the mind in any way." He repeatedly explained how experience leads one to conclude that there are three categories. We skip his argumentation and present only the result. Accordingly, the three categories are *Firstness* —sensation without knowledge of 'what it is,' *Secondness* —our physical reaction to that input, and *Thirdness* —the awareness that there is a pattern

---

[6]https://en.wikipedia.org/wiki/Emergence

that causes that reaction, that the input information has a rule-like character. Since the categories involve experience, thus interaction, they must be dynamic.

Our running example illustrates the three categories. You perceive the visual signal of a red spot on your skin. This signal is "whatever is present to the mind." The input sensation is in the mode of Firstness since you are just 'sensing it.' It is a sensation, a pure feeling without any beginning or end. But then, you become aware that the red spot is 'other to your skin' where it is located and that the two, the red spot and your skin, are parts of the phenomenon you perceived. That state of awareness is in the mode of Secondness since you are aware of the distinctness of the red spot and your skin, that is, that the red spot is 'other-to-you.' It is an experience of 'otherness' (and you experience it; hence, it is affecting you), which you react to because it is other to you, that the red spot is something 'other-to-you.' Then, using your knowledge (your experience or stored Thirdness), you interpret the red spot on your skin as 'mosquito bites.' That comprehension is in the mode of Thirdness. The experience of the distinctness of the red spot and your skin pre-supposes the sensation of different qualities; its meaning as 'mosquito bites' requires that it is tested for that property. Successful testing can contribute to generalization and forming habitual, i.e., rule-like meaning. In the case of mosquito bites, your knowledge base kicks in and tells you what this feeling is and how you should interact with it. Also, your body's biological system kicks in, and your skin swells up as cells try to reject the mosquito venom.

5.3.1 Fundamental Properties

A problem with examples is that we easily get lost in the details. To avoid this danger, we return to an analysis of the properties of the categories. A remarkable feature of the Peircean categories is that they can be applied equally to qualities and thoughts and, thus, in a broader sense, to matter and mind.

The category of Firstness refers to an experience of matter that is purely simple, thus qualitative. This experience is organized within Firstness, which means it is a pure feeling, a sensation without any awareness of where it is coming from or what it is. Firstness is a possibility waiting to be actualized. The second category, Secondness, refers to an experience of matter that is particular, specific, and 'other' to something else. Secondness refers to the organization of matter and thoughts into specific parts, each distinct from the other. Those parts, being 'other' to each other, are organized in relations. Secondness is actualization, waiting to be tested for an abstract concept. The category of Thirdness is also an experience. It is an experience of a rule or a law, the development of habits. These habits do not exist per se, i.e., they do not exist like the specific parts that make up Secondness. They function as 'habits-of-organization' within matter or the rules by which individual members of a collective are generated and organized. Such as DNA. Such as social norms of greeting people. Such as language norms. Thirdness is the capacity for generalization, enabling a meaningful response.[7]

In sum, the categories are ways of classifying experiences in terms of what is present to the mind in any sense, not the physical things that caused the experiences. Contrary to Aristotle's theory, Peircean categories are modes of organization of matter and mind, or types of functioning, not things. In ordinary terms, there are three types of operation: in itself, relative to something else, and in some sense relative to something else.

The three categories are not independent. Categories of a lower ordinal number are involved in higher order categories, but not the reverse. Though Secondness cannot be reduced to Firstness, it presupposes Firstness. Similarly, though Thirdness cannot be reduced to either Firstness or Secondness, it presupposes Firstness (through Secondness) and Secondness, both. Conversely, the categories are related by a relation of subservience. The element of Firstness remains a mere possibility unless some interaction actualizes it, and the one of Secondness remains a brute interaction unless it derives its meaning from Thirdness.

Phenomena can present properties of all three modes of organization. For example, if the visual signal of a red spot on the skin is perceived only as visual quality, it is operational in the mode of Firstness; if it is recognized as a red spot on the arm, it functions in the mode of Secondness; and, if it results in scratching the site of the mosquito bites, it works in the mode of Thirdness. Since phenomena are experienced, categories are classes of information, including learned information, i.e. knowledge. Knowledge must be categorial.

5.3.2   Sign Relations and Categories

What does the Peircean sign concept have to do with the categories? Each of the three sign relations can operate in any of the categories. For example, *quality, actual event,* and *rule* operate in (the mode of) Firstness, Secondness, and Thirdness, respectively. See Fig. 5.4.

Signs are based on categories, but is this relationship mutual? What do Peircean categories have to do with signs? In a word, nothing. More precisely, they share one element: the ability to interpret. Signs arise through interpretation, and interpretation is a distinguishing feature of the third category.

Can we create triadic signs with the computer? The answer is negative. The complexity that characterizes signs and interpretations exceeds the limits of recipe-like processing, let alone the possibilities of computer calculations. What is within the scope of the process model is the creation of input representations analogous to the nine relations. This relation allows it to be used for uniform modeling of

---

[7]This paragraph is based on correspondence with E. Taborsky (personal communication, 2022).

Fig. 5.4 The nine relations and the category of their mode of operation

information processing in different areas of knowledge. However, it cannot express the authentic meaning of ternary signs.

5.4 Pending Issues

Categories and signs are theoretically fundamental, but what is their practical use? Below, we examine several case studies, grouped into two parts, to demonstrate how these concepts interweave human knowledge. The first part deals with the use of categories in concept development. We begin with interpretation as a process and show that its three common understandings are categorially different. Next, we concentrate on syllogistic reasoning, (naive) mathematics, and natural language. Although these domains are fundamentally different, their information is uniformly organized around the Peircean categories. In the second part, we shift our focus to signs, in particular, to symbols, in information processing. We develop a model for syntactic symbol processing and syllogistic reasoning and discuss the possibility of computational implementation.

5.4.1 Three Understandings of Interpretation

Interpretation, as a process, can be distinguished into three types: computational, recipe-like, and meaningful. The first, computational, refers to data processing, including data conveyed by qualities, so the term 'computational' is used in the broadest sense. This concept of interpretation refers to possible meaning, the meaning of the input in itself. An example is sensory information processing which functions in this mode. Computer calculations also fit into this scheme. Since the computer is not related to sensation, how can such calculations be regarded as interpretation? We repeatedly said that the computer is qualitatively less complex than human processing. And yet, we can also speak of (a kind of) experience in the case of the computer, namely in the state transitions of the active elements, such as transistors, that make up its hardware. The second, recipe-like, is formal relational, which works on natural qualities. This concept of interpretation refers to (f)actual meaning expressed by the dependence between perceived qualities (cf. ingredients). The third, meaningful, is sign processing. At this level, interpretation comes down to responding to the input based on knowledge about its qualities and their dependencies.

The three conceptions of interpretation are not independent. Recipe-like processing can be derived from meaningful interpretation by considering dependencies between the input qualities only as formal relations. It can also be seen as computation by considering the input qualities and their dependencies as data and combinatory relations. In this sense, the three types of interpretation as a process can be positioned as a Firstness, Secondness, and Thirdness category phenomenon.[8]

5.4.2 Reasoning

The most refined case of interpretation is undoubtedly inferencing, i.e., concluding by reason. The first systematic description of reasoning is Aristotle's syllogistic theory. A syllogism consists of a major premise, a minor premise, and a conclusion, which are propositions. A proposition consists of a subject and a predicate, connected by the copula 'is'. While the subject refers to an entity in the world, the predicate indicates an attribute or property carried by the subject. In syllogistic,[9] which is a formal logic, the subject and the predicate are abstracted in a logical variable. In a syllogism, the minor premise, the major premise, and the conclusion, as logical statements, function, respectively, as a Firstness, Secondness, and Thirdness. The minor premise describes perceived information, and the major premise states knowledge related to that information. The conclusion converts information from the minor into an answer based on knowledge from the major through mediation by the common term.

5.4.3 Number Conceptualization

The categories are fundamental to the concept of natural numbers. In mathematics, natural numbers are used for counting (e.g., 'one') and ordering (e.g., 'first').

---

[8]Another example is determining the input for interpretation. The occurrence of independent qualities of the input interaction is a Firstness; the relation between those qualities and memory is a Secondness; and the interpretation of the input qualities, based on their memory information, as state and effect a Thirdness.

[9]Syllogistic, in logic, is the formal analysis of logical terms and operators and the structures that make it possible to infer true conclusions from given premises. Developed in its original form by Aristotle in his *Prior Analytics* (*Analytica Priora*) about 350 BCE, syllogistic represents the earliest branch of formal logic. See https://www.britannica.com/topic/syllogistic

Examples of number phenomena are estimation of size on some scale ('large'), comparison of sets ('equally large'), and all provable theories in math.

Natural numbers were first formally defined by the Italian mathematician Giuseppe Peano, who laid down their properties in nine axioms.[10] Peano's first axiom asserts zero ('0') is a natural number. The next seven axioms formulate the properties of the relations equality and successor on natural numbers. The above two groups of axioms are interesting because they are already sufficient to define all natural numbers. Yet Peano introduced a ninth axiom: the axiom of induction. Mathematical induction is necessary for proving that a theorem holds for a set of natural numbers, so the property described by the theorem can be meaningful (in a mathematical sense). The three groups of axioms describe the categories of natural number phenomena, where the first axiom describes them as a Firstness, the second to the eighth axioms as a Secondness, and the ninth axiom as a Thirdness.

The depth to which the concept of natural numbers is related to the Peircean categories becomes clear from the analysis of the principle of mathematical induction. Inductive proofs can be divided into three steps, called 'base case,' 'inductive case,' and 'inductive generalization.' The three steps correspond to the formulation of the general property of a theorem as a phenomenon of Firstness, Secondness, and Thirdness.

To get an impression, we consider the following simple theorem: For any natural number n, the sum of all natural numbers from 0 up to n is $n*(n + 1)/2$.

Let us test if the statement is correct. For example, take $n = 3$. Then the sum of the first n numbers, $1 + 2 + 3$, must equal $3*(3 + 1)/2 = 6$, which is true. The question is: How can we prove that the property of the theorem holds for all values of n? It is where the induction principle comes in handy.

Base case:

For 0, $0*(0 + 1)/2 = 0$ is correct. This step is limited to qualitative information about '0' (zero). It is like an expression of the theorem's meaning as a feeling. In that sense, it is a Firstness.

Inductive case:

We assume the theorem is correct for any k smaller than n; that is, the sum of all natural numbers from 0 until $k-1$ is $(k-1)*(k-1 + 1)/2$, or simply $(k-1)*k/2$. So far, we only state qualitative information about natural numbers. Going from $k-1$ to k makes two changes happen. The first is that the sum, $(k-1)*k/2$, is increased with k by taking the next natural number. This gives $(k-1)*k/2 + k = (k*k- k)/2 + k = (k*k + k)/2 = k*(k + 1)/2$. The second is that the formula of the theorem is applied to value k. This gives $k*(k + 1)/2$. The condition for the theorem is that the results of the two changes are equivalent, which holds trivially. The inductive case is an expression of relational information, therefore, a Secondness.

Inductive generalization:

---

[10] An axiom is a statement everyone believes to be true; in mathematics, an axiom is an unprovable rule accepted as true because it is self-evident or particularly useful.

If the statement holds for the base case and it also holds for the inductive case, then it must be true for all natural numbers. This step uses the brain's potential for generalization. It expresses the theorem as a meaningful property and, therefore, is a Thirdness.

The question of why inductive reasoning is truthful is beyond the scope of Peano's axiomatic theory of natural numbers. There is no doubt that Peano took the truth of inductive reasoning for granted as 'natural.' In addition to induction, mathematical proofs often use deductive reasoning as well. Whereas the conclusion of a deductive argument is inevitable, the truth of the conclusion of an inductive argument is probable. The power of mathematical proofs comes from the principle of induction, which also applies to automated theorem provers. It also shows the limitations of formal mathematics, according to which only formally proven theorems are true. In light of Peirce's theory, the truth of inductive generalization must be the consequence of the rule-like nature of knowledge, including knowledge formulated as a mathematical theorem.

5.4.4   Number Perception

While the concept of natural numbers is obvious, the conditions for number perception need to be clarified. Are we born with the capacity for numbers, or do we acquire them through learning? Surprisingly, according to experimental evidence, number conception is an innate brain ability. That this ability underlies counting and mathematical abstraction is, however, hypothetical. Yet we can conclude from the different meanings of the concept of a multitude in mathematics and elsewhere that the 'naive' number concept must be based on the categories.

Why is it that numbers and counting are not trivial? One possible reason is that number as quality is a product of higher-level brain activity rather than sensory perception. This hypothesis is evidenced by the brain's ability to determine the number of *similar* objects that appear in our perception. In line with our theory that higher-level information, such as thoughts and stored knowledge, originates in memory and not at the sensory level, we assume that number as quality (cf. numerosity) can be modeled as complementary information.

The existence of number-coding neurons in the brain was revealed by neurophysiological research by Nieder and his colleagues (Nieder et al., 2002). Such neurons fire maximally in response to a specific preferred number, correctly indicating various displays in which the cues are not confounded. For example, one such neuron can respond maximally to displays of four items, slightly less to collections of three or five things, and not at all to displays of one or two items. It does not matter whether the displays are equal in perimeter, area, shape, linear arrangement, or density; such neurons only consider numerosity. The number-encoding neurons can recognize the number of similar items from 1 up to 5. However, their signal gets vaguer and vaguer for increasingly larger numbers. Many neurons selectively fire 120 ms after display onset, regardless of the number on the screen, indicating

Fig. 5.5 Desert ants approaching their nest (Wadi Rum, Jordan, 2023)

that the neurons are 'counting' without enumeration. Number-selective neurons are tuned to numerical information and other stimuli, depending on the network in which they are embedded.[11]

Based on these experimental results, we assume that number phenomena can be distinguished into three classes: 'one'—the numerosity of an individual object; 'more'—the numerosity of a countable set of similar objects; 'many'—the numerosity of an uncountable set of such objects. The first ('one') describes any object in itself, thus a Firstness; the second ('more') indicates a collection of objects related by similarity, so a Secondness; and, the third ('many') represents a collection of similar objects as a habitual property, thus a Thirdness. Evidence for the three classes of numerosity in natural language can be found in phrases such as a 'glass,' a 'pair of glasses,' and 'many glasses.' Let us remark that English has no term specific to three, four, or more similar objects.

Hypothetically, the three types of numerosity, 'one,' 'more,' and 'many,' are the basis for counting by accumulation ('one'), via comparison ('more'), and using symbols ('many'). Counting by accumulation, i.e., one by one, is obvious. The ability to order numerosities as more or less, using comparison, is already present in children of age two (Bullock & Gelman, 1977). Counting using number symbols requires learning. Surprisingly, the ability to count by accumulation is already present in animals, as shown by an experimental study in desert ants. See Fig. 5.5.

Under normal circumstances, ants find their way back to their nest based on the direction of the Sun in addition to visual cues, following scent trails. This strategy is not feasible for desert ants because sand can only hold scents short in windy

---

[11] A. Nieder (personal communication, 2012).

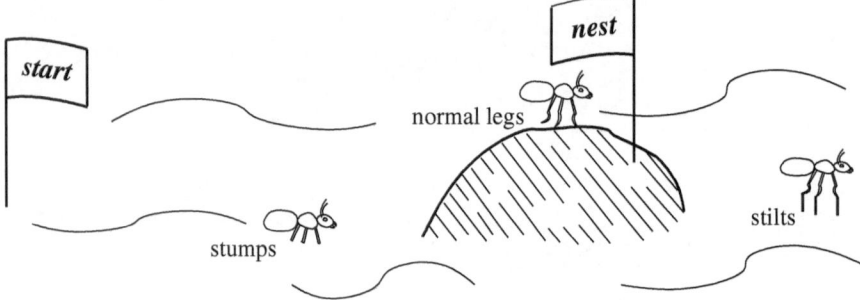

Fig. 5.6 Ants walking from '*start*' towards the assumed position of '*nest*'

conditions. Experimental results show that desert ants measure distances using some step integrator or 'step counter,' thus 'count' by accumulation. In the experiment, the legs, and thus the stride length of the ants, were manipulated. By forcing the ants to move in a tube, walking was limited to a single dimension. Animals with elongated or shortened legs took larger or shorter strides, respectively. Travel distance is overestimated by ants walking on elongated legs and underestimated by ants walking on shortened legs.[12] See Fig. 5.6.

5.4.5   Natural Language

While the significance of numbers is indisputable, the most important domain of knowledge must be natural language. Here, we argue that also language concepts are based on the three Peircean categories.

Historically, the relationship between language and categories was quite the opposite. Aristotle developed his category theory based on his analysis of language, not the other way around. The reason for analyzing language, in particular sentences, besides the development of a theory of truth, was the growing importance of the Greek language in the Hellenistic period, between the death of Alexander the Great (323 BC) and the annexation of Greek heartlands by Rome (146 BC). It became imperative for those who aspired to a career in the civil service, the army, and commerce to have a good command of the Greek language. It created a huge demand for Greek language and grammar education (Seuren, 1998). Aristotle's contributions to grammar were subject, predicate, and case (morphological modification of a word, for example, the derivation of adjectives from nouns). Surprisingly, he did not have a term for the notion of the sentence.

---

[12]https://www.wedrawanimals.com/how-to-draw-an-ant

Below, we will analyze language syntax to illustrate that natural language concepts can be based on Peircean categories. Before that, we briefly touch on traditional language theory, specifically Chomsky's Universal Grammar.

5.4.5.1 Universal Grammar

According to Noam Chomsky, all languages contain similar structures and rules. The fact that children everywhere learn their native language in the same way and without difficulty indicates that we are born with basic principles already present in the brain. We can acquire language because we are genetically encoded with a Universal Grammar, a basic understanding of how human communication is structured.

Chomsky claimed that all languages share three primary structures: lexical category, relation, and phrase. As for their informative content, lexical category, say, a noun, informs about a word in itself; relation refers to a dependency between lexical items, for example, a noun and an adjective modifier; a phrase stands for a combination of lexical items in a structure based on a rule, e.g., the merging of a verb and its arguments into a verb phrase, based on the verb's argument structure.

The connection between the three types of structures and the three categories should be clear. In addition to these primitive structures, recursion (or encapsulation) is another fundamental form of language expression. With rare exceptions, every language has a structure that can repeat itself and thus can, in principle, be extended to an unlimited depth.

Chomsky speculated that Universal Grammar might be extremely simple and abstract, for example, only a mechanism for combining symbols in a particular way. He elaborated his idea in his X-bar theory of syntactic category formation.[13]

5.4.5.2 Cognitive Reality

Psycholinguistic research on adjective-noun combinations provides evidence for category-based organization in natural language (Draskovic et al. 2001). The results indicate that subjects distinguish such combinations into intersective, subsective-compatible, and subsective-incompatible. The analysis below shows that the three types of adjective-noun combinations are Firstness, Secondness, and Thirdness phenomena.

Intersective type adjective-noun combinations denote something that exists on its own. An example is 'yellow car,' referring to an object that is both 'yellow' and is a 'car' (note that 'yellow' is a potential meaning, which is actualized by 'car'). Suppose there are many things around, some of which are 'yellow.' Then the utterance, 'Show me a yellow one,' makes sense.

---

[13] The symbol X denotes a lexical category, X-bar or X' a relation, and X-double bar or X" a phrase.

Subsective-compatible type adjective-noun combinations involve the meaning of a link between two entities. An example is 'interesting car.' There may be several 'cars' nearby, and we can select a subset of them by pointing to certain 'cars' via 'interesting.' The utterance, 'Show me an interesting one,' can only be meaningful when a collection of 'cars' has already been selected. The determination of 'car' is limited by 'interesting,' complementing the meaning of 'car' in the sense of restriction, i.e., negatively.[14]

Subsective-incompatible type adjective-noun combinations involve the meaning of a rule. An example is the phrase 'fast car.' A 'car' can be 'fast' because the meaning of the first involves the element of speed, which can be modified by 'fast' in an intersective or subsective-compatible sense of combination. 'Car' and 'fast' together define the symbolic conventional meaning of 'fast car.'

The findings obtained in the experiments testing the semantic interpretation of the three combinations show differences in computational complexity, with intersective combinations being the simplest and the two subsective types being progressively more and more complex.

We are done with the first part of 'pending issues.' In the second part, we focus on sign processing in natural language and reasoning, including the possibility of computational implementation.

5.4.5.3 What Is 'Natural' in Natural Language?

Language encompasses all sorts of articulation, including texts, facial expressions, and gestures, to mention a few. A common element they all share is a symbolic reference based on a conventional meaning. A language example is the word 'Monday.' That this symbol denotes a day of the week is apparent only to those familiar with this convention. For others, it is only a word, so just a form. The combinatory rules of symbols in language syntax are also conventional. Unless you know them, their meaning remains hidden.

Language phenomena, e.g., a sentence, appear as a series of symbols recognized as one meaning. Although the analogy with apparent motion detection is tempting, we have no evidence. Here, we propose that language processing can be described based on the process model, with events expressing a relationship between linguistic symbols. Language has different levels of abstraction, commonly called morpho-syntax, syntax, and semantics. The first concerns the question of how morphemes contribute to a syntactic symbol. The second and third deal with the problem of how, from words, syntactically and semantically well-formed sentences can be obtained. Language processing at all three levels can be expressed using the process model. In Chap. 6, we will go one step further and suggest that all phenomena, including non-symbolic ones, can be modeled uniformly.

---

[14] according to the meaning of *complementation* as 'naive' logical negation.

What is natural in natural language? The use of categories and signs, and the kind of processing that characterizes all phenomena.

5.4.6 Syntactic Language Modeling

To master language modeling, we focus on syntactic phenomena. But even such phenomena are too complex due to their numerous types, forcing us to lower our ambitions, abandon the idea of introducing a comprehensive language model, and limit ourselves to an illustration of the syntactic processing of simple utterances. However, complexity is not the only problem we face. Since such a thing as 'the rules of language' does not exist, we cannot prove that our model can handle all syntactic phenomena, let alone all language phenomena. Nevertheless, we hope the examples give confidence in a Peircean approach to language processing.

Before going any further, there is something that we should tell about syntactic phenomena. Syntactic rules are concerned with arranging words and phrases into a sentence (syntax is Greek for 'together' and 'sequence'). An example, in English, is the rule that adjectives precede their nominal argument. Remarkably, syntactic phenomena are limited to form. Semantically different symbols, such as 'like' and 'dislike,' can be equivalent syntactically. Syntactic rules can explain a lot, but the meaning of a language phrase or sentence is beyond their possibilities.

5.4.6.1 Syntactic Signs

Syntactic symbols are a type of sign. Not all signs are symbolic, but we already know that very well. The object of a syntactic symbol is the syntactic relation it stands for, e.g., the relation between an adjective and a noun. The interpretant of a syntactic symbol is the sense the relation is formed, which can be neutral, passive, or active. In this way, syntactic phenomena can be classified as Firstness, Secondness, and Thirdness. For a neutral relation, an example is a word occurring in itself; for a passive relation, it is a phrase, such as an adjectival modification of a noun, which is optional. For an active one, it is the subject-predicate relationship, defining the sentence, which is mandatory.

The formal (i.e., form-related) meaning of syntactic signs enables syntactic interpretation to be modeled as a recipe-like process. Syntactic events, establishing a syntactic relation in a neutral, passive, and active sense, can be linked to the events *abstraction*, *complementation*, and *predication*, respectively. See Fig. 5.7.

5.4.6.2 Syntactic Symbol Events

We can consider language phenomena from a linguistic and a process perspective. According to the latter, syntactic phenomena are an interaction between a state and

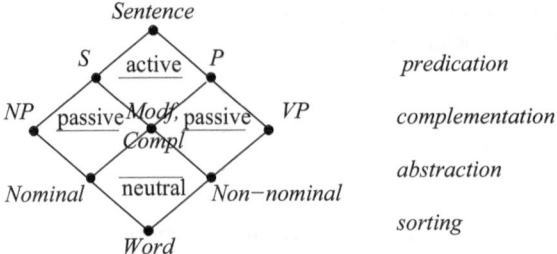

Fig. 5.7 Syntactic symbol relations (*NP/VP* = noun/verb phrase, *S* = subject, *P* = predicate; *Modf* = adjective/adverb modifier, *Compl* = verb complement, *Nominal* = noun, *Non-nominal* = adjective, adverb, etc., *Word* = morpho-syntactically finished syntactic symbol

an effect. The goal of syntactic processing, a syntactically well-formed sentence, is a relationship between the input state as the subject and the input effect as the predicate. Because of the sequential nature of language processing, the relation between the input words, as state and effect meanings, is more complex. A process model of syntactic phenomena can be explained as follows.

In *sorting*, the input words are distinguished into two types, nominal and non-nominal, based on their lexical category. Nominals are nouns (state); non-nominals are adverbs, adjectives, and verbs (effect). In *sorting*, which from a relational perspective is an event that does not occur, the input is further distinguished into symbols that are in focus, as a noun phrase (NP) and a verb phrase (VP), and into complementary symbols, such as an adjective and an adverb, and a nominal which is a verb-complement.[15] Symbols in focus and those that are complementary are distinguished by *abstraction* rather than sorting as a consequence of the sequential nature of the input. In *complementation*, which from a syntactic perspective is optional and, thus, a passive event, the input is represented as a noun-adjective and a verb-adverb modification phrase and as a relation of a verb with its verb-complement (s). A noun and verb phrases, emerging from *complementation*, can function as subject and predicate, respectively. In *predication*, which is obligatory and, thus, an active event, the relation between the syntactic subject and predicate is presented as a sentence.

Sample syntactic utterances are (1) 'fast car,' (2) 'likes very much,' and (3) 'Mary likes gardening.' The used syntactic symbols and symbol relations are (1) 'car' = noun, 'fast' = adjective, 'fast car' = adjective phrase; (2) 'likes' = verb, 'very much' = adverb, 'likes very much' = adverb phrase; (3) 'gardening' = nominal, 'likes' = verb, 'likes gardening' = verb phrase; 'Mary' = subject, 'likes gardening' = predicate, and 'Mary likes gardening' = sentence.

---

[15]Complementary syntactic symbols (adjective, adverb, and nominal) can also designate a phrase.

5.4.6.3 Syntactic Sign Relations

Language phenomena are signs and can be looked at from a process perspective. Appearing input words, the meaning of which in the sentence is, as yet, unknown, function in the mode of *quality*. A nominal symbol that is a syntactic constituent operates in the mode of *likeness*. A noun phrase that can become the subject functions in the mode of *qualitative possibility* and a noun phrase that is the subject in the mode of *actual existence*. An adjective, adverb, and verb symbol, as an appearing property, operate in the mode of an *actual event*. A verb phrase can function in the mode of *rule*; a verb phrase that is a predicate, in the mode of *convention*.[16] The conventional in the syntactic predicate is the agreement that it applies to the subject. Finally, the sentence, which is a statement or proposition, functions in the mode of *argument*.

5.4.6.4 Sequential Processing

Language processing is not trivial, partly because of the sequential nature of language phenomena. The words making up a syntactic phenomenon, such as a sentence, are not immediately available but are presented individually, one by one. Although consecutive input words are related by their order of appearance, they may not be syntactically related.

When the input is available in a single collection, such as in our example of mosquito bites, interpretation can access all information, just like in chess. Information about the input qualia in themselves and their possible relationships with each other is sufficient to determine the meaning of the input as a whole. When the input symbols appear in a sequence, we cannot know whether the input is complete. Therefore, the import of a word or phrase into the sentence cannot be uniquely determined. As a result, language processing suffers from non-determinism.

Yet the sequential nature of the input does not need to be a problem, and the meaning of the input can be smoothly developed during language processing, at least by human agents. Here is an example. Suppose somebody says to you: 'A man came in . . .'. The pause makes you think this should be the whole sentence, with 'a man' as the subject and 'came in' as the predicate. But then the speaker continues: '. . . who was covered with mud.' The additional input makes you reconsider your hypothesis and conclude that the subject must be 'a man who was covered with mud.'

That said, the question is how words contribute to the meaning of the entire input series or, in terms of the process model, how symbols can 'reach' the position in which they realize their syntactic function. How can the process model be used for parsing? To this end, let us consider the properties specific to sequential processing. The first is that consecutively appearing words are independent (*input*). This feature

---

[16] Adjectives morpho-syntactically completed with a copula can also serve as a predicate.

heralds the second property, according to which these independent words cannot refer to the same phenomenon. They are not yet known to contribute to a single phenomenon. Accordingly, expressions of the input[17] as a state (*state-in-input*) and effect (*effect-in-input*) cannot co-occur. The third property is related to *abstraction*. In this event, the input is represented as an abstract concept, depriving it of its meaning as an independent word (word order information may still be available). As a result, all more developed input expressions can already enter into a syntactic relationship with each other and thus represent the meaning of the input series, ultimately as a sentence.

Non-determinism, mentioned earlier, does not restrict itself to the *input* event; it extends to all process events. An oft-cited illustration of this problem is the garden-path sentence:[18] 'The horse raced past the barn fell.' The hypothesis of the predicate needs to be revised in brute processing: not 'raced' but 'fell' is the predicate. Language syntax 'protects' itself against non-determinism in many ways. Information about the lexical category of a word, the number of necessary complements, and word order can make parsing more deterministic and, in this way, increase the efficiency of syntactic processing. An example of such information is the rule that adjectives precede their nominal complements in English, that verbs have a certain number of arguments, and that adjectives and adverbs convey complementary information regarding the input as a sentence.

5.4.6.5 Syntactic Parsing Examples

We illustrate syntactic parsing with a couple of examples. Information available to the parser when processing an input word is depicted by a 'diamond' structure. Events changing the state of the parser are indicated with an arrow. Since, in sequential processing, we cannot determine whether the input is complete or whether more symbols will come, we assume that a constant number of dot symbols terminate the input series of words.[19]

Finally, there is the issue of efficiency. In short, an input word can have different syntactic meanings, and the question is which one the parsing process should realize first. For example, a word can function only as a noun, a noun phrase, or even as the subject of a sentence. We can represent a noun immediately as the subject. However, a noun can be modified later, and a noun phrase can become a verb complement later. In short, we cannot determine whether an input representation is optimal. To make parsing robust against revisions due to non-determinism, we assume that the interpretation is conservative and always selects the least developed meaning of a symbol among the possible ones. The dependency between the syntactic concepts

---

[17] In sequential processing, the input collection (*input*) consists of a single quality.

[18] https://en.wikipedia.org/wiki/Garden-path_sentence

[19] Adding a constant number of dots to the sequence of input symbols (equal to the number of positions in the process model, i.e., nine) keeps the process's computational complexity linear.

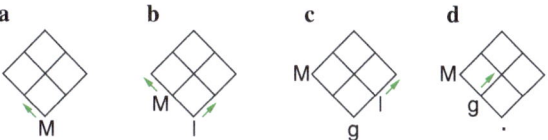

Fig. 5.8 'Mary likes gardening' (Mary = 'M,' likes = 'l,' gardening = 'g,' dot = '.'). A 'diamond' denotes the state of the parsing process; an arrow indicates a change in the parser's state

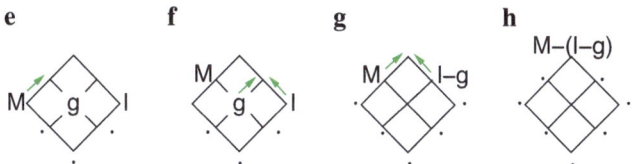

Fig. 5.9 Syntactic parsing (cont)

also justifies this approach: a subject presupposes a noun phrase, which needs a nominal.

The first example is the utterance, 'Mary likes gardening.' Syntactic processing develops as follows. The appearing first symbol, 'Mary,' is interpreted as a word. See Fig. 5.8a. For the same reason, the next symbol, 'likes,' must be interpreted as a word. Since 'Mary' and 'likes,' as words (cf. qualities) are independent, they differ. Their difference, appearing as an effect, forces the parser to rethink its assumption about 'Mary' functioning as a word. As a result, it assigns 'Mary' the more advanced meaning of nominal constituent. See Fig. 5.8b.

Analogously, 'gardening,' which appears as a word, forces 'likes' to be represented as a non-nominal based on the verb category of the latter. Since 'Mary' and 'likes' cannot co-occur in their current meaning, 'Mary' is assigned the more developed meaning of a noun phrase. See Fig. 5.8c. No more words are waiting, but analysis to terminate has yet to reach its goal: interpreting the entire input series as a well-formed sentence. It is when dot symbols, enforcing pending symbol relations to established, prove their usefulness. The first dot symbol forces 'gardening' to be interpreted as a nominal constituent and, subsequently, 'likes' to be represented as a verb. The latter hypothesis is justified by 'likes' being a transitive verb and having a rule-like syntactic argument structure. See Fig. 5.8d.

The next dot symbol forces the constituent 'gardening' to be interpreted as complementary information. This hypothesis is reinforced by the failure of the alternative (which is omitted), where 'gardening' is assumed to be the syntactic subject. See Fig. 5.9e. Since 'Mary' and 'gardening' cannot establish a syntactic modification relation (cf. complementation), the appearing dot symbol forces 'Mary' to be represented as the syntactic subject. See Fig. 5.9f. The subsequent dot symbol makes 'likes,' completed by 'gardening,' to be represented as the predicate, 'likes-

Fig. 5.10 Syntactic
meanings collected

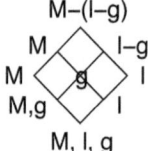

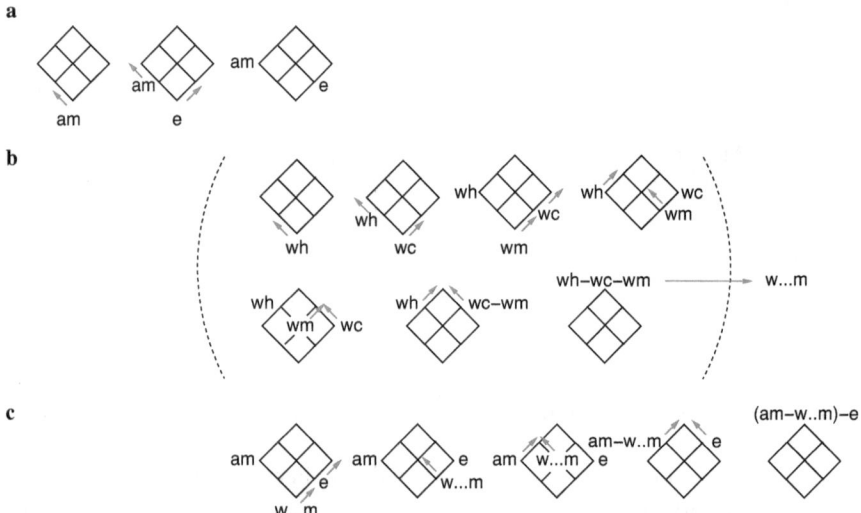

Fig. 5.11 Syntactic parsing of 'A man entered who was covered with mud' ('a man' = am, 'entered' = e, 'who' = wh, 'was covered' = wc, 'with mud' = wm)

gardening.' Note that word-order information from these symbols complies with English's SV(O) rule.[20] See Fig. 5.9g. Finally, the relation of the subject and predicate, representing the entire input series as a sentence, is expressed by a single symbol: 'Mary-(likes-gardening)' See Fig. 5.9h. The entire set of recognized syntactic concepts is shown in Fig. 5.10.

The second example is the utterance we briefly discussed earlier, 'A man entered who was covered with mud.' That the syntactic subject is 'a man who was covered with mud,' and the predicate 'entered' should be clear. This example is exciting because the predicate appears before the subject is finished, seemingly contradicting English's SV(O) rule. Human agents can handle this problem easily. Figure 5.11 illustrates that the process model can solve this problem just as elegantly. A morpho-syntactic analysis can precede syntactic analysis. For the current example, this means that 'a man,' 'was covered,' and 'with mud' are treated as individual syntactic symbols (cf. words).

---

[20] Subject-Verb(-Object).

Major premise: S_1 is P_1
Minor premise: $\Rightarrow$ $\dfrac{S_2 \text{ is } P_2}{S_2 \text{ is } P_1}$ $\{S_1 = P_2\}$
Conclusion:

Fig. 5.12 The syllogistic scheme figure-1. The common term is indicated by $\{S_1 = P_2\}$

Returning to syntactic analysis, the first part of the input, 'a man entered', is parsed as usual. See Fig. 5.11a. This process stops abruptly at the appearance of the symbol 'who,' which marks the beginning of a nested phenomenon. An analysis of this second part, 'who was covered with mud,' is depicted in Fig. 5.11b. The arising input expression, denoted by wh-wc-wm (in short w...m), is used only as an input quality in the analysis of the encompassing syntactic clause. See Fig. 5.11c.

As opposed to syntactic knowledge, which is homogeneous, semantic information can be distinguished into many sub-domains. While this may seem like making language processing too complex, it is not. Language symbols function as signs in all domains and therefore, can be processed uniformly. Uniform representation enables efficient processing by combining information from symbols that function in an identical mode (same sign relationship) in different domains through structural coordination.

5.4.7 The Process Model of Reasoning

Another example of sequential processing is syllogistic reasoning. Why would reasoning be sequential? Propositions are statements concerning individual phenomena. Their meaning can be processed as usual; sequential processing is not a prerequisite. While this is true for the premises, it does not hold for the conclusion. Since inferencing needs information from both the major and minor premises, their interpretation must precede the processing of the conclusion,[21] which sets additional conditions for reasoning as a process. Introducing a process model of reasoning also has an unexpected result. Although syllogistic reasoning is deductive, it also involves, in a sense, the meaning of inductive and hypothetical inference.

Syllogisms can be classified into three figures based on the common term that appears in both premises but not in the conclusion. In addition to the three figures, Aristotle introduced four modes of quantification, which we shall return to in the next chapter. Here, we focus on his figure-1, from which the other two figures can be derived via transformation. See Fig. 5.12.

Previously, in Chap. 3, we mentioned that a statement can be perceived as a premise. This time, the focus is on reasoning as a process. Before proceeding with a process model of figure-1, let us delve into how memory can represent propositional

---

[21] The premises must be treated first as full-scale (triadic) signs and only then as a quality in the processing of the conclusion. This condition may be the ground for logical reasoning presupposing conscious thinking, as suggested in Chap. 3.

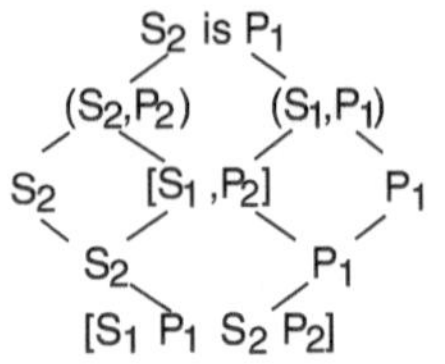

Fig. 5.13 The syllogistic
scheme figure-1 as a process
(we use abbreviations: [S_1,
P_2] = 'context'; (S_1,
P_1) = 'S_1 completed with
P_1')

$$S_2 \text{ is } P_1$$
$$(S_2,P_2) \qquad (S_1,P_1)$$
$$S_2 \qquad [S_1,P_2] \qquad P_1$$
$$S_2 \qquad\qquad P_1$$
$$[S_1\ P_1\ S_2\ P_2]$$

information. We assume that limitations in storage capacity prevent the brain from storing ternary relations such as propositions in memory. A practical argument, Yarbus's experimental evidence, suggests that an external stimulus can only transform memory information into a meaningful sign. This result supports the hypothesis that memory is not a store of ready-made ternary signs. In line with this idea, we assume that propositions are stored as binary relations, not triadic ones. In this way, memory knowledge denotes a possible meaning, which can become (f)actual through interpretation. Regarding reasoning as a process, memory information about a proposition, 'S is P,' is available only as a relation (S, P) and ultimately as a pair of interconnected independent qualities [S, P].

Let us now return to figure-1. As a relation (S, P), a proposition can be viewed from the point of view of both S and P. This possibility is also the key to modeling syllogistic reasoning as a process. We assume that the input is defined by the major and minor premises, as relations. While the minor premise (S_2, P_2) is viewed as a state in focus (S_2), which is related to a complementary property or effect (P_2), the major premise (S_1, P_1) is seen as an effect in focus (P_1), which is related to a complementary state (S_1) that it affects. In summary, the input for syllogistic reasoning can be defined as a collection of qualities: [$S_2\ P_1\ S_1\ P_2$].

From here, a process model of figure-1 can be defined, as usual. See Fig. 5.13. *Sorting* and *abstraction* need no explanation. *Complementation* has two instances that are different from a logical point of view. Regarding S_2 (*state*), we assume that information from S_1 mediated by P_2 (*context*) is used to extend the meaning of S_2 (with factual data about the properties of S_1) and represent it as a statement of an 'actually' existent state (S_2, P_2). Since this information is already known from the minor premise and is thus only recapitulated, this event has the meaning of *deductive* inference. As for P_1 (*effect*), we assume that information from P_2 mediated by S_1 is used to adjust the meaning of P_1. By testing P_1 for P_2 as an instance 'known as S_1', P_1 is positioned as a conventional relation (S_1, P_1). Since testing can generalize P_1's meaning, this event has the character of *inductive* inference. Although information of (S_1, P_1) is already known from the major premise, testing and optional generalization (see (S_1, P_1)) and stating factual information (see (S_2, P_2)) function in different modes. Finally, in the event of *predication*, the relationship between (S_2, P_2) and (S_1, P_1) is expressed by the proposition 'S_2 is P_1 in the context of [S_1, P_2]'; in short, 'S_2 is P_1'. Since the emerging proposition can explain the input only from a single perspective, marked by the common term ([S_1, P_2]), this event has the meaning of hypothetical inference, also known as *retroduction*.

Major premise:	S_1 is P_1		S_1 is P_1	
Minor premise:	$\Rightarrow$ S_2 is P_2	$\{P_1 = P_2\}$	$\Rightarrow$ S_2 is P_2	$\{S_1 = S_2\}$
Conclusion:	S_2 is S_1		P_2 is P_1	

Fig. 5.14 The syllogistic schemes figure-2 (left) and figure-3 (right)

How are the three kinds of reasoning implemented? The answer lies in the way context information (*context*) functions. Since the context refers to S_1 and P_2 as synonymous meanings, context information can be used in two ways. Remember that, following our theory, effects can be represented as a dense domain of values that is a linear ordering and states as an average of a data collection. Based on this idea, *complementation* as a deductive inference can be realized by adjusting the average value of S_2 by values from S_1, thus by information from P_2; and, analogously, *complementation* as an inductive inference can be implemented by using P_2, and so S_1, as a measure in the dense domain of P_1. An example of the latter case is motion velocity as a domain, adjusted by an observed speed value as a (velocity) measure. Finally, the hypothetical inference, by *predication*, can be realized by representing the emerging proposition 'S_2 is P_1' as a relation (S_2, P_1) offered for memorization.

The model of the figure-1 shows that reasoning, as a process, is no different from any other phenomenon. To define such a process is a collection of qualities, a set of dependencies between those qualities, and a goal of interpretation is needed. The model of figure-1 also reveals that interpretation, as a process, involves the meaning of all three kinds of logical reasoning: deduction, induction, and retroduction. Seen from a broader perspective, this means that all phenomena can be interpreted as reasoning. However, this may not explain why such a process would reach the conscious realm.

Syllogistic reasoning has two more schemes, figure-2 and figure-3, in addition to figure-1. See Fig. 5.14. How about the processing of these schemes? We assume that the process model only applies to figure-1. Processing the other two figures requires their schemes first to be transformed into figure-1. This transformation boils down to converting the major premise in figure-2 and the minor premise in figure-3 (Dumitriu, 1977). What we need to know about the three schemes is that the dependence between their terms corresponds to the information structure of the three cases of logical reasoning as a method. In sum, figure-1 involves the meaning of deduction, figure-3 the meaning of induction, and figure-2 the meaning of retroduction (cf. hypothetical inference).

Establishing the converse of a propositional relation may not be simple. Differences in conversion complexity may explain the differences in truth perception in the three kinds of reasoning, where deductive reasoning is trivially true (no conversion), inductive reasoning is less obvious (conversion of the minor premise), and retroductive reasoning is only hypothetically true (conversion of the major premise). That deductive reasoning is valid goes without saying. The possibility of a different truth perception in the other two cases of logical reasoning can be explained as follows.

In reasoning, as a process, information from premises is treated identically, only relational data. The difference between the major and minor premise follows from their import in the syllogistic inference. The major premise is a statement of a general nature. In contrast, the minor premise refers to a particular case related to the subject of the major. Conversion transforms a proposition, 'S is P', into 'P is S.' For the major premise, this can be implemented by collecting all information about S and P into a set, selecting an ordered subset from S's data as a new predicate term (P), and determining an average value representation of P's data as a new subject term (S), and finally by treating the relation between the new subject and predicate as a major premise. The result must be compatible with the minor premise. All in all, it can be a laborious operation. In the case of the minor premise, which is about a particular observation, conversion means visiting elements of the data set represented by the minor premise and introducing a new subject (S) and predicate (P) until the meaning of their relation is compatible with the major premise.

The difference in achieving conversion implies that figure-2 (retroduction) can be more demanding in terms of complexity than figure-3 (induction), which could explain the difference in their truth perception.

5.4.8   Computational Modeling

So far, we have focused on a recipe-like model of information processing. From here, a computational model is just a few steps away. If the assumption of a single type of process holds, a computational model can be defined uniformly for all knowledge domains. The nine relations can be represented as data types and the dependency between them as a procedure. In the following, we limit ourselves to the conditions of a computational model of syntactic information processing.

What is required for a computational parsing model is a definition of input symbols as combinatory needs and a specification of symbol relations as operations on the combinatory needs of their constituents. Nouns can be assigned a passive combinatorial need, in line with their role as an argument in syntactic modification or the subject in predication relation, and adjectives, adverbs, prepositions, and verbs, an active combinatory need, based on their function as a modifier or the predicate. Establishing a syntactic relation means resolving constituent symbols' corresponding passive and active combinatory needs.[22] Symbols that cannot establish a syntactic relation with a type of syntactic symbol are assigned a neutral combinatory need specific to that type. For example, in English, adjectives are relationally neutral to verbs following them in the input (i.e., at surface level).

---

[22]Combining identical types of combinatory needs through accumulation also belongs to the possibilities.

The combinatory need of a symbol can be specified in any detail. An example is a verb, which can be neutral (for an adjective),[23] passive (for an adverb), and active (for verb arguments). The last one can be further refined into intransitive (no verb complements), transitive (single verb complement), and di-transitive (two verb complements). For example, 'give <>,' 'give <a book>,' and 'give <Mary><a book>.'

While the specification of combinatory needs of symbols (cf. lexicon) can be cumbersome, the definition of a process model using such a lexicon can be straight-forward. The resulting parser can be surprisingly efficient due to the algorithmic complexity of the process model, which is linear in the number of input symbols and thus less complex than context-free parsing. Linear complexity means that each input symbol can be processed by executing a constant number of instructions on a reference Turing machine; in other words, the number of instructions needed to process a single symbol is independent of the number of input symbols (cf. input length). Linearly complex algorithms allow for efficient computational implementation.

What about the non-determinism of syntactic parsing? Does it not destroy the property of linear complexity? Hypothesis revision requires the parser to take a 'step back' and consider other alternatives. Book-keeping of parsing decisions, known from backtracking parsers, can be more complex and make the parsing process take longer. However, since the size of the lexicon is limited by a constant, the linear complexity of parsing holds invariantly.

How difficult is it to create a lexical definition of a language, such as a syntactic lexicon? The possibility of characterizing syntactic relations as phenomena enables a lexical definition to be developed systematically. While creating such a specification can be labor intensive, it only needs to be done once (cf. compile-time), so it does not affect the complexity of the parsing process (cf. run-time).

---

[23] Post-modifying adjectives are an exception.

Chapter 6
Knowledge Is Categorial

Abstract Traditionally, interpretation refers to human processing. This book aims to show that this concept can be applied more broadly. The connection between structural properties of the model of meaningful processing of this book on the one hand and fundamental natural phenomena, spherical waves, on the other, suggests that all of nature must be capable of interpretation: knowledge must be inevitable. What effect might this conclusion have on the ongoing technological revolution? While traditional artificial intelligence does not impose conditions on computational realization, in other words, it can use any program, we argue that meaningful processing can be based on one type of process. The ultimate goal of a human-like thinking computer can be achieved by using the classes of sign relations as data types (and specifying knowledge in terms of the categories) and using the dependencies between the classes of sign relations to model information processing as a procedure.

6.1 The Nature of Knowledge

What you see as this chapter's title is also the book's conclusion. It follows syllogistically from the titles of previous chapters, as the attentive reader may have noticed. See Fig. 6.1. As for our goal, the reasoning employed therein must be reversed. After all, categories form the basis for signs, and according to a theory of cognitive activity, signs form the basis of our knowledge.

It looks good, but we can only be partially satisfied with this result for two reasons. First, we have yet to verify that the process model is compatible with human interpretation. To be sure, we need experimental data from as many knowledge domains as possible. Our evidence so far is limited to syntactic language processing and reasoning. Although the results show that the concepts used by the subjects are compatible with the representations according to the process model, at least in the knowledge domains examined, more data is needed for a firm answer. The second reason concerns the rationale behind interpretation as a process. Is it an invention of the brain, or is it rooted in nature? It is this question we will now explore in more detail.

© The Author(s), under exclusive license to Springer Nature Switzerland AG 2025   111
J. J. Sarbo, *Inevitable Knowledge*, SpringerBriefs in Computer Science,
https://doi.org/10.1007/978-3-031-73461-8_6

Chapter 2	Knowledge	is	(formed by) Cognition
Chapter 3	Cognition	is	(formed by) Brain processes
→	**Knowledge**	**is**	**(formed by) Brain processes**
Chapter 4	Brain processes are		(formed by) Sign processes
→	**Knowledge**	**is**	**(formed by) Sign processes**
Chapter 5	Sign processes are		(formed by) Categories
	Knowledge	is	(formed by) Categories
Chapter 6	≈ **Knowledge is categorial**		

Fig. 6.1 Syllogistic conclusion of the book

6.1.1 Interpreting Nature

It is believed that, in a broader sense, animals and plants are also capable of interpretation in addition to humans. Could interpretation be a characteristic of all phenomena, thus nature? What does it say about the relationship between matter and mind?

Peirce was unequivocal on this point[1] by stating that "matter is effete mind, inveterate habits becoming physical laws." Neither is primary, neither is derived from the other, and neither is separate. Peirce accepted that material objects exist outside what we think of them and do not arise from the mind. He rejected materialism, which recognizes that the mind is evolved from matter.

The suggestion that matter and mind are intimately linked is fascinating. Unfortunately, this idea, like the concept of a ternary sign relation (a.k.a. Peircean sign type), is not immediately useful for our purposes. Proving that nature is an interpreting system is beyond our limits. However, it is within our reach to demonstrate that fundamental physical phenomena such as spherical waves can, in some sense, express the process meaning of interpretation.

A characteristic of spherical waves is the possibility of interference. An interaction between waves is itself a wave. The analogy with the concepts of potential sign (cf. wave interaction) and interpreted sign (cf. reaction wave) should be clear. We will show that wave phenomena can express the process model's events geometrically as a development, starting with the input as quality, expressed as a point, and ending with the input as a proposition, expressed as a sphere.[2] The interpretation moments by the process arise as geometric structures on the reaction wavefront. An unexpected result is that those structures are isomorphic to the Platonic solids.

---

[1] C.S. Peirce (1932), vol. 6., paragraph 24.

[2] This analysis is based on research initiated by J.I. Farkas. See (Farkas & Sarbo, 2018).

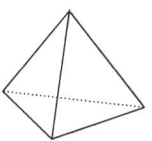 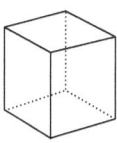 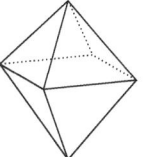 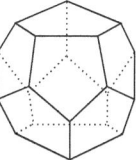 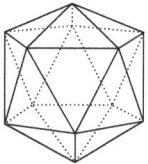

Fig. 6.2 The five regular polyhedra, also known as Platonic solids: the tetrahedron, cube, octahedron, dodecahedron, and icosahedron

6.1.2 Platonic Solids

In three-dimensional space, a Platonic solid is a regular, convex polyhedron. 'Polyhedron' means it is made by joining polygonal faces; 'convex' means having an outline or surface curved like the exterior of a circle or sphere; 'regular' means that all angles are equal in the faces, and all sides are equal. Polygonal faces of a Platonic solid are identical in shape and size and meet the same number of faces at each vertex. The classical result is that only five convex regular polyhedra exist.[3] These are depicted in Fig. 6.2. The five solids that meet the criteria are the tetrahedron, cube, octahedron, dodecahedron, and icosahedron. Each of them has a dual Platonic solid with faces and vertices interchanged. The tetrahedron is self-dual, the cube is dual with the octahedron, and the dodecahedron is dual with the icosahedron. In their constitution, the dodecahedron and icosahedron involve the golden ratio.[4]

6.1.3 Wave Phenomena

Wave phenomena are known for two specific properties: their homogeneous character and the Huygens-Fresnel principle. The first, which we already mentioned, means that a wave interaction is itself a wave phenomenon (cf. reaction wave). According to the second, any unobstructed point on the wavefront is the source of spherical wavelets. Secondary wavelets emanating from different points mutually interfere, and their sum forms a new wavefront. The original and new wavefronts convey the same information, analogous to the process model, representing the input as a collection of qualities (original wave) and a proposition (reaction wave).

We need to look at the process model from a novel perspective to grasp the geometric meaning of wave phenomena. So far, we have focused on the possibility of embedding the process model in the hierarchy of sign relations. This time, we will concentrate on the sign relations, particularly their dependency, not between them, which we already know and will use in the current analysis, but internally. We aim to assign a geometric meaning to those internal dependencies of the sign relations.

---

[3] https://www.mathsisfun.com/geometry/platonic-solids-why-five.html

[4] https://en.wikipedia.org/wiki/Golden_ratio

Establishing a sign relation can also be modeled as a process. It will not surprise us that the geometric and the processual views of interpretation can be comparable. Based on this link, we will show that sign relations can be characterized as matching and analysis mode processes, besides their meaning as a geometrical structure. The connection between the two kinds of meanings, the geometric and the processual, is, however, limited to analogy; a complete definition of sign relations as a process is beyond our purpose. We only note that such a process can be incomplete, meaning that one or more of the input representations are combined into a single expression or, geometrically, represented as a single vertex. Similar to previous applications of the process model, the initial and final expressions by a sign relation's process are referred to as potential and propositional meanings.

Below, we go through the nine sign relations individually, revealing their dependence structure and geometric representation. We must admit that the below analysis is labor-intensive. Readers interested only in the main ideas can skip to Sect. 6.1.5, which presents an overview of the results.

For our analysis, we need a whole set of abbreviations. We refer to the interacting waves as h_1 and h_2 and the reaction wave as h_3. The three waves are independent analogously to other interaction phenomena. We assume that h_1 affects h_2 (an analysis of the opposite case, which is isomorphic, h_2 affects h_1, is omitted). Waves are indicated by symbols in italics (e.g., h_1); as quality, they are given in plain text (e.g., h_1 is denoted as quality by h_1). Orthogonal axes are used to represent independent waves (and qualities). Edges between a pair of vertices designate a relation. Geometric structures are denoted by symbols in italics, e.g., *point*. Int = interaction, st = state, eff = effect, co-ex = co-existence, and co-occ = co-occurrence. The location of the input wave interaction is referred to as origo (O).

With so many new symbols, it is not unreasonable to reiterate what it is all about. In short, the input wave interaction is between h_1 and h_2, and the reaction wave is h_3. Our goal is to show that the wavefront of h_3 is a carrier of structures that geometrically represent the nine sign relations.

Quality   The reaction wave, h_3, also as quality, conveys information about h_1 and h_2. Since the three waves are independent, their meaning as qualities and unary relations (h_1, h_2, and h_3) are isomorphic and hence can be merged. Geometric expression: *point*.[5]

Actual event   Because the independent waves, h_1, h_2, and h_3, appear simultaneously, the input interaction trivially involves the meaning of co-occurrence. This dependence between the three waves can be geometrically represented by three points of mutually perpendicular axes, h_1, h_2, and h_3 (the last one as h_{1*2}). Co-occurrence relation is expressed from the point of view of the qualities, individually, through edges. For example, h_1 is related to h_2 and h_{1*2}. See Fig. 6.3a. Geometric expression: (equilateral) *triangle*.

---

[5]The point, square, and equilateral triangle are considered regular polyhedra (degenerate, in a geometric sense).

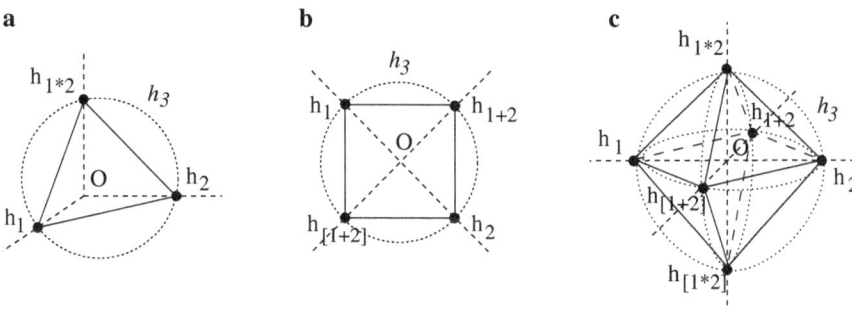

Fig. 6.3 Sign relations as geometric structures. (**a**) Actual event; (**b**) Likeness; (**c**) Connection. '$*$' = logical and, '$+$' = logical or; $h_{[1 + 2]}$ and $h_{1 + 2}$ are a logical or- relation of h_1 and h_2 as potential and propositional meaning; idem for $h_{[1*2]}$ and h_{1*2} as a logical and-relation of h_1 and h_2. 'O' = origo

Likeness The input wave interaction involves the meaning of co-existence in h_1 appearing with h_2, h_2 appearing with h_1, and h_1 and h_2 appearing with each other. Co-existence is about a relation between constituents, that is, selection without separation (note that co-occurrence has no analogous meaning). A possible case of selection is when neither h_1 nor h_2 is selected, when the input interaction is recognized only as an existence relation, thus as co-existence with itself, without identifying the interacting qualities as constituents. In sum, input co-existence can be represented by four different terms: 'neither' ($h_{[1 + 2]}$), 'one' (h_1), 'other' (h_2), and 'both' ($h_{1 + 2}$), which define an induced partial ordering of information where 'neither' is the most simple and 'both' the most developed. This ordering, thus input co-existence, can be represented by antipodal points on a pair of orthogonal axes. The dependency between the meanings of co-existence can be understood as a matching mode process: $h_{[1 + 2]}$ (int), h_1 (st), h_2 (eff), and $h_{1 + 2}$ (co-ex), where $h_{[1 + 2]}$ and $h_{1 + 2}$ represents co-existence of h_1 and h_2 as potential and propositional meaning. See Fig. 6.3b. Geometric expression: *square*.

Connection As an interpreting system, the reaction wave (h_3) merges the two types of input information, focus and complementary,[6] into a single meaning. To this end, complementary information about input co-existence and co-occurrence is expressed as synonymous relations, according to our model. We are particularly interested in the geometrical meaning of this dependence between the two types of complementary data. Synonymous expression requires that co-existence is regarded as co-occurrence, and vice-versa, co-occurrence as co-existence (everything only in the sense of a binary relation). The possibility of such a relationship follows from the two meanings of h_1 and h_2: a constituent in itself and a participant in an actual event, which are related, and even more clearly from the meaning of $h_{1 + 2}$ and h_{1*2}, which

---

[6]Theoretically, distinguishing between the two types of qualities is possible through Fourier analysis. Qualities in focus and those complementary can be represented by the first n ($n \geq 0$) and all higher harmonics, respectively.

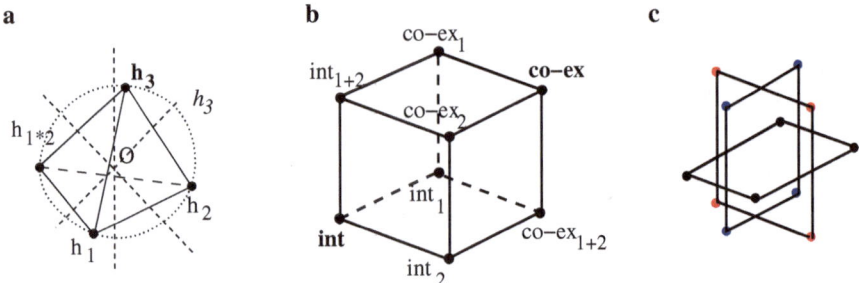

Fig. 6.4 Sign relations as geometric structures (cont.). (**a**) Rule; (**b**) Qualitative possibility; (**c**) parallel faces of the cube merged into golden squares. In (**b**), abstract co-existence as potential meaning is represented by **int**, int_1, int_2, and int_{1+2}, and as propositional meaning by **co-ex**, $co-ex_1$, $co-ex_2$, and $co-ex_{1+2}$ (cf. the meaning of co-existence from different perspectives). We use bold notation when a symbol refers to an unusual meaning. These are **int** and **co-ex**, as representations shared by various structures (and processes), and h_3, as rule-like input meaning

differ basically in the aspect of co-occurrence, a meaning which is involved in the second but not in the first.[7] An example of a synonymous representation of co-existence and co-occurrence is the dependency defined by h_1 and $h_{[1+2]}$ (co-existence meanings appearing simultaneously) and $h_{[1*2]}$ (co-occurrence as a potential meaning). Note, that all quality symbols above, e.g., h_1, h_2, $h_{[1+2]}$, etc., refer to complementary information. See Fig. 6.3c. Geometric expression: (regular) *octahedron*.

Rule We have seen that input co-occurrence can be represented by binary relations (edges) between h_1, h_2, and h_3 as h_{1*2}. In the case of a rule-like input meaning, these binary relations are abstracted into a ternary relation. Such a relationship calls for a new type of vertex: $\mathbf{h_3}$. The new vertex is linked to h_1, h_2, and h_{1*2} as vertices and represents their triadic relationship. See Fig. 6.4a. Geometric expression: (regular) *tetrahedron*.

Qualitative possibility Input co-existence can be abstracted from its three perspectives, 'one,' 'other,' and 'both,' and therein, as potential and propositional meaning ('neither' allows for an expression of abstract co-existence only in the sense of existence). In sum, there are six possible outcomes of abstraction that we elaborate on one; the others can be developed analogously. In keeping with the meaning of abstraction, we use formal symbols to express abstract co-existence. The expression we delve into in more detail is when we take the perspective of 'one' and consider input co-existence as potential meaning. In this case, we only have the input collection of qualities as an existence; we denote it by **int** ('neither'). One of the meanings of co-existence is the occurrence of one of the input qualities ('one'), which we formally refer to by int_1. Another meaning of co-existence is the occurrence of the other input quality. Viewed the input and, thus, the other input quality

---

[7] h_{1+2} refers to both h_1 and h_2 as constituents; h_{1*2} designates h_1 and h_2 as co-occurrence

figure-1	**mood-1**	**mood-2**	**mood-3**	**mood-4**
C B	*All C B*	*No C B*	*All C B*	*No C B*
A C	*All A C*	*All A C*	*Some A C*	*Some A C*
A B	*All A B*	*No A B*	*Some A B*	*Some A not-B*

Fig. 6.5 The syllogistic scheme figure-1 and its four conclusive moods

from the perspective of int_1, this other quality occurs with int_1 and, therefore, is referred to as int_{1+2} ('other'). The two expressions, int_1 and int_{1+2}, jointly define abstract input co-existence from the point of view of int_1: co-ex$_1$ ('both').

The dependency between **int**, int_1, int_{1+2}, and co-ex$_1$ can be seen as a matching mode process. Combining the six expressions of abstract co-existence, we get the meaning of the input wave interaction as a qualitative possibility. Since int_1 and co-ex$_1$ refer to h_1, and int_2 and co-ex$_2$ to h_2, and finally, int_{1+2} and co-ex$_{1+2}$ to h_3, square faces of the resulting single structure that are parallel refer to an identical representational perspective, i.e., the perspective of h_1, h_2, and h_3 (as h_{1+2}). See also Fig. 6.4b. Geometric expression: *cube*.

Actual existence The input wave interaction as actual existence is a harder nut to crack. A solution can be found in the ability of phenomena to express syllogism in a sense. Syllogisms are arguments about entities, or existing 'things' (subject), characterized by some property (predicate). Aristotle introduced three syllogistic schemes: figure-1, figure-2, and figure-3. In addition, he captured the meaning of figure-1 in four conclusive moods. See Fig. 6.5. Below, we will show that wave phenomena can geometrically express the four moods by a single structure. To this end, we revise the definition of the input wave interaction as a collection of qualities. Since h_1 and h_2 are phenomena, thus an interaction between a state (st) and an effect (eff), the input qualities can be defined as $h_1 = (st_1, eff_1)$ and $h_2 = (st_2, eff_2)$.[8] As a result, the input wave interaction can be conceived as st_1 and eff_2 occurring in the context of st_2 and eff_1 or, in syllogistic terms: $A = st_1$, $B = eff_2$, $C = (st_2, eff_1)$, with C functioning as a common term.

Aristotle's figure-1 consists of three propositions, functioning as major premise ('C is B'), minor premise ('A is C'), and conclusion ('A is B'). Syllogistic is concerned with concluding from premises using deduction. The feasibility of this operation implies that some familiarity with the premises is already available to the reasoning agent. In line with this idea, we assume that the premises describe an interaction between st_1 and eff_1 (major premise) and between st_2 and eff_2 (minor premise) represented as matching mode processes and that the syllogistic combination of the premises as an interaction between st_1 and eff_2 (conclusion) is represented as an analysis mode process. See Fig. 6.6a. The four processes depicted in this diagram correspond to the four moods of figure-1. We assume that quantification is 'ALL' when a proposition is a statement about qualities from a single phenomenon, and 'SOME' and 'NO' when it refers to qualities from different phenomena, which

---

[8]Note the analogy with reasoning (h_1 and h_2 as premises, and h_3 as the conclusion).

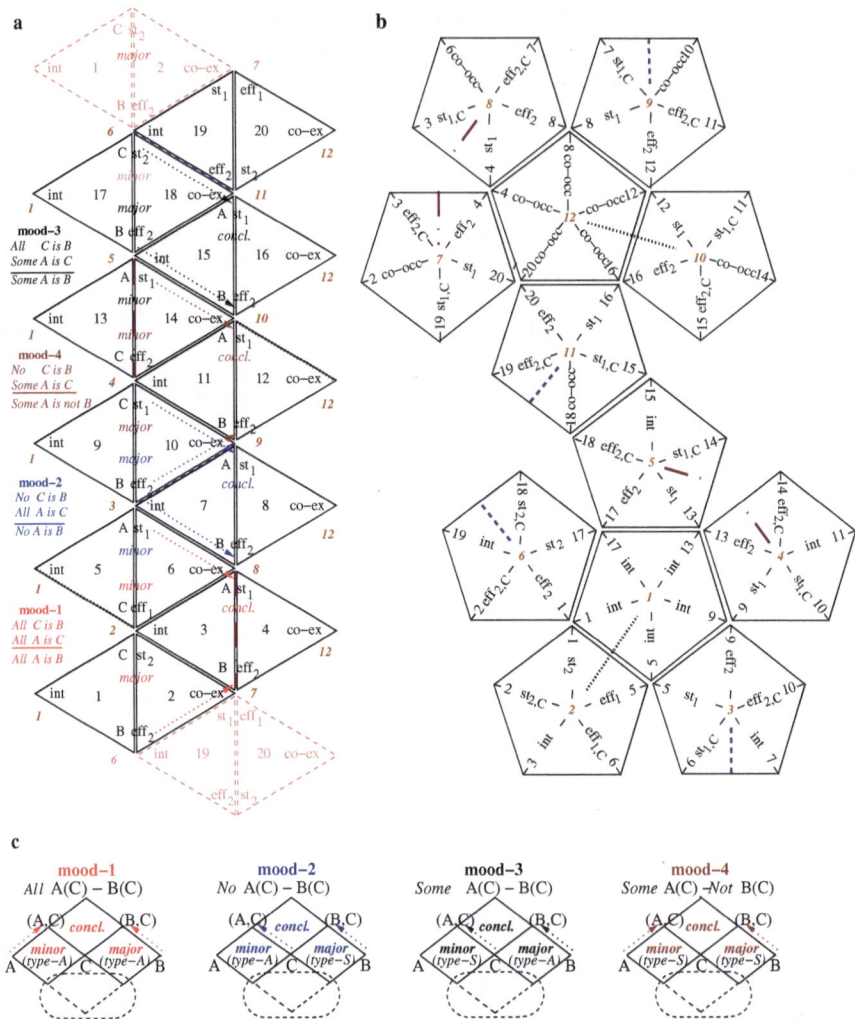

Fig. 6.6 Sign relations as geometric structures (cont.). (**a**) Actual existence. Input qualities are denoted by syllogistic variables, e.g., A = st$_1$ ('either'), B = eff$_2$ ('other'), C = (st$_1$,eff$_2$), or (st$_2$, eff$_1$) ('both'). Triangles representing the four conclusive moods are: (mood-1) 1, ..., 6, (mood-2) 5, ..., 10, (mood-4) 9, ..., 14, (mood-3) 13, ..., 18. Dotted arrows indicate information flow. Symbols assigned to the same vertex denote synonymous meanings. All instances of **int** stand for the same single vertex (the need for using separate vertices is a consequence of planar representation); this also holds for occurrences of **co-ex** labels. (**b**) Convention. We use abbreviations, e.g., st$_1$, $_C$ is short for (st$_1$,C), or context information about st$_1$. In (**a**) and (**b**), golden rectangles are indicated by colored edges with a pattern. (**c**) The four conclusive moods of figure-1 depicted as incomplete analysis mode processes. Interpretation moments that are merged are indicated by a dashed oval. In (**a**), the positions of mood-2 and mood-3 as a process are transposed (see the reverse order of the major and minor premise). Black bullets marking a vertex are omitted

are dependent and independent, respectively ('SOME' can only apply to the subject term, thus to a state type quality). We call the first a Type-A ('ALL') and the last two a Type-S proposition ('SOME' and 'NO').

Recall that each of the four moods of figure-1 is represented by an interaction between st_1 and eff_2 in the context of st_2 and eff_1. In Fig. 6.6a, an example of propositions as a premise quantified as 'ALL' are the matching mode processes completing eff_2 with st_2 and st_1 with eff_1, indicated by the triangles 1-2 and 5-6. Their syllogistic conclusion concerns a single phenomenon, representing the interaction between st_1 and eff_2, as an analysis mode process, indicated by the triangles 1-2-3-4-5-6.

An example of a proposition as a premise quantified by 'NO' is the matching mode process 13–14. In this process, eff_2 (cf. context) refers to st_1, which is meaningless since h_1 and h_2 are independent; thus, their relationship is not yet known[9] (it will only be known after the input has been recognized as a proposition). Finally, an example of a proposition as a premise quantified as 'SOME' is the process 17–18, in which eff_2 (context) completes st_2, which is correct and looks so far to be a Type-A proposition. However, the labels of vertex 5 (context) tell that information by eff_2 is restricted to st_1.[10] In other words, only qualities shared by st_1 and eff_2 are used for completion; thus, 'some' of st_1. The four moods are represented as formal relations. Accordingly, the propositions of h_1 and h_2 (and their processes) can function differently in the four analysis mode processes. In mood-1 and mood-4, it is the expression of the state and the effect of a syllogism as a process; in mood-2 and mood-3, the reverse is the case. In other words, the place of the major and minor premises can be different.

The structure in Fig. 6.6a can be systematically derived by combining the meanings of the wave interaction as a qualitatively possible and as a connection. For the geometric case, this requires joining the six squares of the cube and furnishing their matching mode processes with complementary information. Due to the number of cases, this derivation cannot be set out briefly. We skip details and focus on the big picture. Remember that co-occurrence has three meanings, as opposed to co-existence, which has four. Since parallel faces of the cube refer to the same perspective (one of h_1, h_2, and h_3), their squares can be combined. Based on the meaning of merging as co-occurrence, the three parallel squares as processes are enriched with complementary information in the sense of co-occurrence in three ways, according to their perspective. In addition, the four interpretation moments of the individual processes are completed with the four meanings of complementary information about input coexistence: 'neither' or no information, 'one' or information about h_1 (state), 'other' or information about h_2 (effect), and 'both' or information about h_1 and h_2.

An unexpected consequence of merging parallel faces is that the arising squares are no longer incident to the wavefront of h_3. Their structure must be transformed to

---

[9] It could have been processed in matching mode if the input was known.

[10] Symbols assigned to the same vertex denote a synonymous meaning.

have this feature again, which is when George P. Odom's ingenious construction (Odom and Craats 1986) comes in handy. It shows that by stretching parallel squares in the direction of their shared dimension (one of h_1, h_2, and h_3) and transforming them into golden rectangles, the new corner points will be incident to h_3. See Fig. 6.4c.

Equilateral triangles defined by the (12) vertices of the three golden rectangles represent co-existence from different perspectives. Squares comprised of a pair of triangles can be interpreted as a matching mode process, and three squares linked by a common vertex (context) as an analysis mode process. In the latter case, the resulting process is not complete (it lacks the event of *sorting*); its propositional meaning as an interaction between h_1 (state) and h_2 (effect) is indeterminate. In this way, we end up with four incomplete analysis mode processes. See Fig. 6.6a. Geometric expression: (regular) *icosahedron*.

Convention An ultimate challenge is to explain the input wave interaction as conventionally meaningful. We have seen that a geometric representation of the input state, such as the square, cube, and icosahedron, can be defined in a systematic fashion based on a geometric expression of less developed sign relations, for example, the cube as a structure composed of squares; the icosahedron as a combination of triangles and squares. We do not have an analog definition of the effect. One possible reason for this is the oft-cited property that an effect presupposes the existence of a state, which means that the two meanings cannot be treated separately. A simultaneous construction of both exceeds the possibilities of geometry. We had a similar problem earlier with the rule-like meaning of the wave interaction, and we found a solution by introducing a ternary relation and a new type of vertex: $\mathbf{h_3}$.

We come across the same problem of definition in the case of the conventional meaning of the input effect. Remember that, in *complementation*, the rule-like meaning of the effect is furnished with context information. Since this information is relational, this event can be seen as extending the effect's meaning with additional relations. A geometrical representation of the result is based on this same idea. In Fig. 6.4a, $\mathbf{h_3}$, denoting the input rule-like meaning, is linked with h_{1*2} (co-occ), h_1 (st), and h_2 (eff). From a geometrical perspective, the event of *complementation* introduces new relations as vertices and edges. In Fig. 6.6b, these are indicated by the edges labeled $st_{2,C}$, and $eff_{1,C}$. (e.g., $st_{2,C}$ is st_2 completed by C or information from the context). The arising vertices[11] can be arranged in a pentagon structure, with $\mathbf{h_3}$ in the middle (label omitted).

How is context information used in *complementation*? From the icosahedron, as a set of analysis mode processes, we already know that information about the input state and input effect as co-existent constituents has four types. The effect's co-occurrence meaning dictates that *complementation* considers all possible combinations of co-existence information. Accordingly, co-occurrence can be assigned six

---

[11] representing independent meanings.

different interpretations[12] representing the conventional meaning of the effect as a potential for (**int**) and in the sense of a proposition (**co-occ**). Ultimately, two sets of six pentagonal structures can be introduced this way. The pentagon structures (2*6) are duals of the 4*3 vertices of the icosahedron (cf. four analysis mode processes). While the processes of the icosahedron represent the possible meanings of the subject of the input interaction, the pentagons express the different meanings of the predicate. Like in the case of the subject, the pentagon structures of the effect as a process are incomplete, this time due to the lack of an expression of the *predication* event.

How can *complementation* be represented geometrically? In this regard, we mention that the vertices of the octahedron define six squares (four on the surface, two on the inside). Their data can be interpreted as coexistence information from a particular perspective. Data from the six squares of the octahedron can be used to complete information by the six squares of the cube. A similar correspondence is found between the eight triangles on the surface of the octahedron on the one hand, and the four vertices of the tetrahedron as ternary relations and the four triangles on the surface of the tetrahedron as a binary relational expression of co-occurrence on the other. It looks promising, number-wise, but that is all we can say about it.

Returning to the icosahedron as four incomplete analysis mode processes, if *sorting* is not represented, and so information about the input as quality is not available, then the meaning of the input as a syllogistic inference can only be known in the sense of structure (cf. subject). Regarding the pentagon structures of the effect, the situation is different. The lack of an expression of the *predication* event implies that the input is not known as a proposition, only as a property (cf. predicate). See Fig. 6.6b. Geometric expression: (regular) *dodecahedron*.

Proposition The dependency between the interaction as actual existence and as conventional property is geometrically expressed by the duality of the icosahedron and the dodecahedron. By viewing the hierarchy of sign relations as a process, *predication* combines the two kinds of information about the input interaction into a single meaning.[13]

In the case of wave phenomena, such a meaning can be attributed to the spherical wavefront defined by the emerging and interfering wavelets, according to the Huygens-Fresnel principle. Accordingly, interpretation can be characterized geometrically as transforming the input wave interaction that appears as a point into a wavefront with point-like properties and which (response) wave can function as a quality in a subsequent wave interaction. Geometric expression: *sphere*.[14]

---

[12]Formally, the number of combinations equals '2 under 4', that is, $4!/(2!*2!) = 24/(2*2) = 6$.

[13]This event, which applies to the effect and state, cannot be defined geometrically and for the same reasons as in the case of the complementation of the input rule-like effect.

[14]The sphere falls beyond the realm of Platonic solids.

Fig. 6.7 Tetrahedron
(black), the involved
octahedron (red), and
therein, a cube (blue)

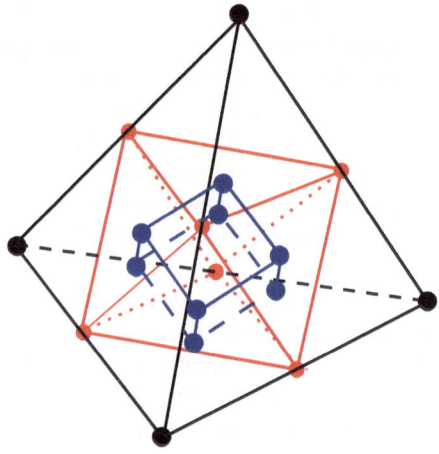

6.1.4   Remarks

Although we cannot systematically derive a geometric representation of the conventional meaning of an input effect, the geometric perspective used is powerful enough to express a particular feature of the effect we have so often referred to: the assumption of an involved state by the effect. See Fig. 6.7. This diagram shows that the tetrahedron contains an octahedron and, within that, a cube. Vertices of the cube fall on the midpoints of the faces of the octahedron. This interdependence between these three structures indicates that, in the *complementation* event, faces of the cube (cf. involved state) could be transformed into golden rectangles and, via mediation by the octahedron (cf. context), vertices of the arising polygon used as centers of the 2*6 pentagons representing the input effect as a predicate.

It is worth noting that a matching mode process can express the meaning of co-existence and co-occurrence. However, while the first is complete (see square), the second is not (see triangle). The reason for this difference lies in the ability of a state to exist on its own (so complete), but an effect cannot (so incomplete). The geometric expressions of the effect indicate the state involved in three ways: by co-occurrence as quality, in the case of the triangle; by a relationship, in the case of the tetrahedron; and by a meaningful geometrical property, the duality with the icosahedron, in the case of the dodecahedron. Finally, we remark that, as a dependency, sign relations that are more complex than *connection* (cf. context) can be modeled as an analysis mode process; all others can be positioned as a matching mode process.

6.1.5   An Overview of the Results

A remarkable result of the above geometric analysis is that Platonic solids, thus also the golden ratio as a property (traditionally denoted by the Greek letter φ), are involved in wave phenomena. Just as the sphere is beyond the Platonic realm, so is the proposition beyond the domain of Boolean logic and the sentence falls outside the theory of language syntax. Finally, let us note that duality as a phenomenon, functions as a Firstness in the self-duality of the tetrahedron and the self-consistency of rule-like meaning, as a Secondness in the relationship between the cube and the octahedron and between the input interaction as a qualitatively possible and a connection (relation); and as a Thirdness, in the duality of the icosa- and the dodecahedron, and the dual meanings of the input as an actual existence and a conventional property.

Figure 6.8 gives an overview of the geometric interpretation. Since all phenomena have wave-like properties, all phenomena, thus nature, must have a capacity for interpretation in some sense.[15] Knowledge must be inevitable.

6.2   Charles Sanders Peirce (1839–1914)

Peirce biographer Joseph Brent writes: "Peirce was a polymath, at home in the physical sciences, especially chemistry, geodesy, metrology, and astronomy. He was [...] a master of logic and mathematics, an inventor of the field of semeiotics.[16] In philosophy, he was one of the most original thinkers and system builders of any time, and indeed, the greatest philosopher the United States has ever seen."

Brent is not alone in his admiration. Peirce's achievements are similarly recognized by Bertrand Russell, Umberto Eco, Karl Popper, and Noam Chomsky. Remarkably, Peirce was one of the first to prove the existence of an operator

Fig. 6.8 A geometric interpretation of the nine relations

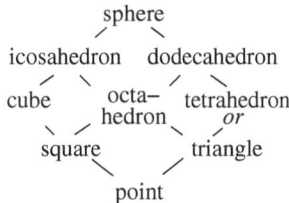

---

[15]Is memorization natural? Surprisingly, it can be. Evidence shows that vanadium dioxide can 'remember' like a human. In this material, electrical activation causes a structural effect (Nikoo et al., 2022). Note the analogy with the growth of new axons due to learning.

[16]Semeiotic(s) is Peirce's doctrine of signs.

sufficient for all logical functions.[17] His logical-nor operator ('not or') is fundamental to switching circuits comprising today's computer hardware. The relevance of his research is undoubtedly due primarily to his theory of categories and signs, which are unparalleled in their originality.

What is the importance of Peirce's ideas to the current digital age, especially to Artificial Intelligence (AI)? For a long time, the increasing complexity of data was seen as a threat to information processing. Limitations of current technology are evidenced by the fact that despite the tremendous computational power and storage capacity, no computer system can meet the complexity of human intelligence. Recent AI applications that use deep learning[18] (learning structural properties from data) seem to make a breakthrough possible. However, more than information processing based on statistics is needed to meet the condition of being meaningful. The result of computational learning may be wrong if the training data is not characteristic of a particular phenomenon. Due to the formal nature of computation, justifying the output as meaningful may not be obvious. The problem of computers that can think human-like is more sophisticated than expected. Using a single type of process and uniform data representation enabled by Peircean theory provides better opportunities to solve that problem.

Aristotelian theory dominated philosophy and science for more than two millennia. Peirce's concept of a dynamic phenomenon is more fundamental and paradigmatically new. Only a few such turns are known in history. One of them is, of course, Aristotle's entity-property theory itself. Another one is Newtonian mechanics.[19] In particular, Newton's assumption that motion[20] is a state, the change of which requires a force, contrary to the Aristotelian view that motion is a process that needs a force.[21] Finally, the celestial theories of Copernicus and Kepler, which we briefly discussed, also fit into this list.

Peirce is one of the foremost scientists in history. That he had the ambition himself to become influential is witnessed by his wish "to make a philosophy like that of Aristotle, that is to say, to outline a theory so comprehensive that, for a long time to come, the entire work of human reason, in philosophy, of every school and kind, in mathematics, in psychology, in physical science, in history, in sociology, and in whatever other department there may be, shall appear as the filling up of its details," as he puts it in one of his articles (Peirce 1887).

---

[17] Peirce proved the functional completeness of the logical-nor operator in 1880. Henry M. Sheffer introduced a similar operator, the logical-nand ('not and'), in 1913.

[18] The neural network Minerva of Google has trillions of hidden variables. Besides high energy consumption, this implies long learning time and low robustness (noise in the input can cause failures in recognition). In its simplest form, a neural network comprises an input, hidden, and output layer. In deep learning, the hidden layer includes several levels of hidden neural networks defining a structure (Ananthaswamy, 2023).

[19] Newton built his theory on Galilei's theory of relativity (Hecht, 2015).

[20] motion at a constant velocity.

[21] According to everyday experience, things in motion tend to stop.

There are only a few books about Peirce's life and character. Historian Max Harold Fisch was appointed official biographer a year after the last of those who had a personal acquaintance with Peirce died. An extended biography, based on original documents and correspondence, is given by Joseph Brent. His dissertation, later published in book form, is also the primary source of information for the next section.

6.2.1 Short Biography

Charles Sanders Peirce was born in Cambridge (MA) on September 10, 1839. Already at a young age, it turned out that he was exceptionally talented. His father, Benjamin Peirce, who insistently considered him a genius, spent long hours with him teaching mathematics and revealing general principles of science. At 25, Charles was invited to give the famous Lowell lectures. A few years later, in 1867, he was elected a resident fellow of the American Academy of Arts and Sciences, of which his father had been president. In 1871, Charles Peirce sailed to Europe for an expedition to observe the solar eclipse. In Great Britain, he visited Augustus De Morgan, a leading mathematician at that time. Peirce's research made a strong impression on Britain's finest logicians and philosophers, and his scientific reputation was increasing in stature. He was at the zenith of his life at this time.

The first dark clouds gathered above him around the same time. One of Charles Peirce's instructors was C.W. Eliot, who disliked him. In 1869, Eliot became president of Harvard College. Within a year, Peirce and Eliot conflicted with Charles's position at Harvard and other matters. Benjamin Peirce was the director of the U.S. Coast Survey. In this function, in 1871, he appointed his son, over the heads of the more experienced assistants, as an acting assistant in charge of gravimetry experiments. Charles Peirce knew very little about geodesy and gravimetric research at this time. His unexpected appointment aroused jealousy and resentment.

In 1877, Charles Peirce was considering the likelihood of an offer by Johns Hopkins, a newly founded research university, to join the faculty, but he only got a lectureship. That it was not a professorship he hoped for is thanks to Eliot, one of the most powerful men in American higher education at that time. Despite all his problems, Peirce continued to research and achieved significant results. He had an extraordinary track record in logic and many other fields of science. In Europe, his reputation was well-established. However, the American Academy professionals could not appreciate Peirce's originality of mind and his true achievements.

His father's death in 1880 deeply affected Charles, personally and professionally. The destruction of his career marked the next few years. He was fired from Johns Hopkins, charged with neglecting his Coast Survey responsibilities, and eventually forced to resign. A person most vicious towards Peirce was Simon Newcomb, his former colleague at Johns Hopkins and presumed friend. Newcomb was a self-made man who became an influential figure in science education. He resented Charles Peirce for his privileged background and his extraordinary talent. He made Charles' reputation a subject of gossip at the academy and the Survey. All this undoubtedly contributed to Peirce's misfortune: he was never offered a regular academic job again. Newcomb was the editor of the American Journal of Mathematics. In this position, he declined a vital paper by Charles Peirce, on the ground that it was not mathematics. Arguably, Newcomb lacked the competence to make such a judgment. A few years later, when Charles Peirce applied for a grant, it was again Newcomb's denial of the value of Peirce's work that prevented him from receiving the funding.

Peirce was on the verge of bankruptcy due to the lack of a regular income. It was partly a result of his refusal to adapt his lifestyle to his income. In the years after 1893, Charles Peirce's fortunes reached their lowest level. From 1905, he lived in deep poverty and had virtually no income. His health also deteriorated rapidly. Still, he diligently continued his research until he could not work. Between 1884 and when he left Johns Hopkins, until his death, he produced some of his most valuable results. In 1909, he was diagnosed with abdominal cancer. Charles Peirce died on April 19, 1914.

6.2.2   Fate and Destiny

What may explain Peirce's misfortune? Unfortunately, since there is so little evidence about his life, any answer can only be speculative. However, Brent's account makes it possible to track down some of the circumstances that led to Peirce's doom, which has the hallmarks of a Greek tragedy: the downfall of a hero who is from a higher rank (in Peirce's case, his high intelligence) and who commits a mistake because of his pride, and that mistake leads to a disaster.

Charles Peirce was born into an upper-middle-class family. He was considered clever and handsome. By all means, he had everything a man could want. Yet, he became an outcast and lived in extreme poverty at the end of his life. How could this happen?

Charles' father, Benjamin Peirce, was a professor at one of the best U.-S. universities and held various public positions. He was a famous mathematician, but he was not one of the greatest of his time. Today, he is best known as the 'father of.' If he was ambitious, as we can reasonably assume from his academic and public career, his belief in his son's extraordinary talent may have been driven by the hope that he would make the breakthrough he failed to achieve. How else can we explain his insistence on seeing his son as a genius? If Benjamin Peirce was such a person, he must have known the heavy burden that role would put on his son's shoulders. According to Brent (1993)[22] one of the effects of his father training him for genius was to provoke Charles, "an arrogance toward and impatience with others who exhibited a superficial understanding of lack of clarity in their thinking." Speculatively, we can assume that his father's constant belief that his son was a genius must

---

[22]pg. 57.

have come to Charles as a mission to realize. Children indoctrinated with an ideal, positive or negative, ensure they live up to it. It can make a child depressed but also narcissistic, among others, if he believes to be a genius indeed. There is no evidence that Peirce was such a child. However, in his life, he was not accepted by academics and had conflicts with influential people. His persistence to research, even in times of illness and poverty, can be seen as evidence of living up to a mission.

As Brent puts it, his father's high expectations made him exceptionally competitive. Competitiveness is often accompanied by a permanent sense of alertness and an inability to relax and make friends. Peirce had very few friends in his life, indeed. Proud, if only for his social background and unusual talent, he constantly resisted any authority except that of his father and chose his own goals that rarely met the demands of others. Peirce believed that his results should convince everyone. No doubt this is why, time and again, he solicited a job and asked for financial support from people who disliked him and did not even look into his publications.

His troubled relationship with others may explain why he could not recognize Newcomb's hatred and jealousy toward him, even after he had been his colleague at Johns Hopkins for 4 years. Peirce believed his value to humanity was significant enough to impose on others. According to Brent, he could not understand his responsibility for his destruction. He could find no reason for his failure except the cruelty and error of others. Despite his incredible talent, he consistently failed in practice. History shows that he was not alone in this problem. In the seventeenth century, the French scientist Blaise Pascal wondered why mathematicians could do so poorly in everyday life. In his notebook 'Pensées,'[23] he writes it must be the case that axiomatic thinking does not apply to practical problems. A lack of sufficient hypotheses can lead to insufferable decisions. While Pascal's diagnosis sounds plausible, there is another possible explanation, which is the ignoring of social and emotional cues. It is not unlikely Peirce had such blindness. Had he noticed those signals, he would have recognized their importance and acted accordingly.

6.2.3 The Secret of Genius

What does it take to become a genius? Is this a matter of especially high intelligence? Or, it is enough to be above-average smart, and, ultimately, it is a matter of luck (which is, in essence, Daniel Kahneman's idea of success).[24] While the first answer is plausible, the second sounds just as real. Luck is necessary for success because all revolutionary progress begins with some simple premises and induction, and the question is which premises to take. The prerequisite for finding structure in the abundance of information we are confronted with daily is to have a workable hypothesis. It has a lot in common with a chicken and egg situation. Premises

---

[23] "The difference between the mathematical mind and the practical mind" in B. Pascal (2013).

[24] https://alearningaday.blog/2019/02/08/daniel-kahnemans-success-equations

function as glasses to the world, but premises are only a start. Since theory development can take a long time, the premises must be known at the youngest age possible. How lucky are geniuses in this regard? A few historical examples give some insight. By helping his father, the physician Nicomachus, treat patients, Aristotle learned the importance of observation before the age of thirteen. Newton learned about the properties of mechanical force by creating working models at the age of ten. By understanding the principles of logic at the age of twelve, Peirce became acquainted with the formation of arguments.[25]

In everyday life, we have to make many decisions, and it happens that we make a wrong one. That this happens to the ordinary person is not unusual. That it can also occur to a genius is more unexpected. Again, historical examples show that this is possible. Early in his career, Aristotle ironically criticized a powerful aristocratic party enemy, Isocrates, one of the most influential Greek rhetoricians of his time, without regard to the consequences. Eventually, Aristotle had to flee Athens. Aristotle's teacher, Plato, had it worse. Plato claimed that philosophers would be the best kind of rulers. Fascinated by the perfection of his idea, he ignored the problems of reigning. King Dionysius II of Syracuse (Sicily, Italy) allowed him to put his theory into practice. Eventually, it resulted in chaos. According to the historian Diogenes Laërtius, Plato so much angered Dionysius that he sold him into slavery as a punishment. Favored by fortune, Plato was ransomed shortly later.[26] It is known that Newton was not without errors either. He was an esteemed scientist in England and was appointed the president of the Royal Society. In a few years, his lust for power turned him into a dictator who would not tolerate criticism. He brutally opposed Leibniz, accusing him of plagiarizing his differential calculus, which was utterly wrong. The irony of history is that Leibniz's formalism, gaining in elegance, has been used by science ever since. In his old age, Newton developed a bitterly suspicious personality. He made many enemies and was feared and hated by many.[27]

Brent writes: "The builder, almost creator of Peirce's character, was not fate; it was his father." In the lack of evidence, we can only speculate the reasons for Peirce's tragic fate. According to Brent, Peirce's big mistake was his arrogant behavior. However, Brent never met Peirce or spoke to anyone who knew him, so his conclusion should be taken with a grain of salt. Peirce must have felt an injustice in the constant rejection of his requests. The behavior of his enemies cannot be rationally justified. It was inhuman. Not only because of the poverty it created but also because of being brutally disrespectful. In the words of Willem James (Peirce's

---

[25] In his autobiographical notes, Einstein writes: "After ten years of reflection, such a principle [i.e., relativity] resulted from a paradox upon which I had already hit at the age of sixteen." https://en.wikipedia.org/wiki/Einstein%27s_thought_experiments

[26] One source of the story of Plato being sold into slavery is "Lives and Opinions of Eminent Philosophers" by Diogenes Laërtius. https://history.stackexchange.com/questions/37495/was-plato-sold-into-slavery

[27] The only people he loved were his (only) friend and his young niece. In his last years, Newton lived a solitary life.

best friend), "The deepest principle of human nature is the craving to be appreciated."

Peirce made numerous discoveries, many of which have also been rediscovered by others in the past hundred years. But no one came up with an idea that could rival the originality of his concept of signs and categories. He first published his categories very young, at the age of twenty-seven, and he was aware of the great significance of his discovery. According to Brent (1993),[28] in his maturity, Peirce said, "The theory of the categories [...] is the gift I make to the world. That is my child. In it I shall live when oblivion has me – my body." Admittedly, his visionary thinking is not characterized by modesty, yet he was right.

The importance of Peirce's category theory is justified, among others, by his theory of signs, which is also based on categories. Peirce argued that a sign only functions as a sign if it is interpreted as such. By looking at signs from this angle, i.e., from a processing perspective, the functioning brain[29] can be seen as the Representamen (or the functioning interpreting system). In light of this, human processing can be seen as a transformation from the input stimulus, which appears as a quality (cf. matter), to brain activity (cf. mind) and finally into a response (cf. matter). If interpretation and mind apply to all of nature, Peirce's conjecture must be correct: Matter becomes mind, and mind becomes matter.[30]

6.3 Epilogue

We provide an overview of a theory of meaningful processing. In retrospect, the underlying research was guided by intuition rather than a preconceived plan. Since the conclusions are known by now, the likely premises can be deduced by reasoning backward. It shows that the premises and the research questions were not arbitrary. They functioned as markers in the process of comprehending Peirce's theory.

---

[28] pg. 70.

[29] The functioning brain and the mind are considered to be identical.

[30] Matter and mind remain different but, only work in combination (Taborsky, 2003), (Dilworth, 2016).

References

Ankersmit, F. R. Sublime Historical Experience, Stanford University Press (2005) https://www.journals.uchicago.edu/doi/10.1086/517549

Aristotle: In J. Barnes (ed.), Complete Works of Aristotle, Princeton University Press, (1984)

Ananthaswamy, A.: In AI is bigger better, Nature, vol. 615 (2023)

Barsalou, L.W.: Flexibility, structure, and linguistic vagary: Manifestations of a compositional system of perceptual symbols. In: A.C. Collins, S.E. Gathercole, M.A. Conway, & P.E. Morris (eds.), Theories of memory, pp. 29-101, Lawrence Erlbaum Associates, London (1993)

Bense, M.: Das System der Theoretischen Semiotik. Semiosis, no. 1, pp. 24-28 (1976)

Bense, M., Walther, E.: Wörterbuch der Semiotik. Kiepenheuer & Witsch, Köln (1973)

Brent, J., Charles Sanders Peirce: A Life, Indiana University Press, Bloomington and Indianapolis (1993, 1998)

Broadbent, D.B.: The magic number seven after fifteen years. In: A. Kennedy and A. Wilkes (eds.), Studies in long term memory, Wiley, London, pp. 1-18 (1975)

Broussard, M.: More than a Glitch, The MIT Press (2023)

Bullock M., Gelman, R.: Numerical reasoning in young children: The ordering principle, Child Development, vol. 48, no. 2, pp. 427-434 (1977)

Chai, W.J., et al.: Working Memory from the psychological and neurosciences perspectives: A review, Front. Psychol.; Rockefeller University (2018) doi:10.3389/fpsyg.2018.00401. A revised map of where working memory resides in the brain. ScienceDaily, 29 September (2020) www.sciencedaily.com/releases/2020/09/200929123352.htm

Chomsky, N.: Knowledge of language: its nature, origin and use, Praeger, New York (1986)\

Chomsky, N.: Syntactic structures, The Hague, Mouton (1957)

Churchland, P.: The Big Questions: Do we have free will? New Scientist, issue 2578, pp. 43-45 (2006)

Cook, V., Newson, M.: Chomsky's Universal Grammar: An Introduction (3rd ed.). Wiley-Blackwell (2007)

Davey, B.A., Priestley, H.A.: Introduction to lattices and order, Cambridge University Press (1990)

Dehaene, S.: Single-Neuron Arithmetic, Science, 297, pp. 1652-1653 (2002)

DeWall, C.N., Baumeister, R.F., Masicampo, E.J.: Evidence that logical reasoning depends on conscious processing, Consciousness and Cognition, 17, pp. 628-645 (2008)

Dilworth, D.A.: Peirce's Transmutation of Schelling's *Philosophie der Natur*, *Cognitio*, 17 (2), pp. 253-90 (2016)

Draskovic, I., Pustejovsky, J., Schreuder, Adjective-Noun Composition and the Generative Lexicon. In: P. Bouillon and K. Kanzaki (eds.), Proc. of the First Int'l Workshop on Generative Approaches to the Lexicon, Universite de Geneve (2001)

Draskovic, I., Couwenberg, M., Sarbo, J.J.: New Concept Development Model: Explorative study of its usability in Intelligence Augmentation. In: J. Cordeiro (ed.), Proc. 2nd Int'l Conf. on Computer Supported Education (CSEDU 2010), pp. 288-292 (2010)

Dumitriu, A.: History of Logic, vol. 1-4, Abacus, Tunbridge Wells (1977)

Farkas, J.I., Sarbo, J.J.: Interpreting nature, Cognitive Systems Research, vol. 49, pp. 1-9 (2018)

Feng-Hsiung, H.: Behind Deep Blue: Building the Computer that Defeated the World Chess Champion, Princeton University Press (2004).

Ferrucci, D., Levas, A., Bagchi, S., Gondek, D., Mueller, E.T.: Watson: Beyond Jeopardy!, Artificial Intelligence 199, pp. 93-105 (2013)

Gazzaniga, M.S.: Cognitive neuroscience, W.W. Norton, New York, London (1998)

Gazzaniga, M.S., Ivry, R.B., Mangun, G. R.: Cognitive neuroscience (2nd ed.), W.W. Norton Company, Inc. (2002)

Gibson, James J.: The Theory of Affordances. The Ecological Approach to Visual Perception. Taylor & Francis, Boulder, Colorado, pp. 119-137 (1979)

Goldstone, R., and Barsalou, L.: Reuniting perception and conception, Cognition, vol. 65, pp. 231-262 (1998)

Hagoort, P., Hald, L., Bastiaansen, M., & Petersson, K. M.: Integration of word meaning and world knowledge in language comprehension. Science, 304(5669), pp. 438-444 (2004)

Harnad, S.: Categorical Perception: The groundwork of cognition, Cambridge University Press, Cambridge, UK (1987)

Hecht, E.: Origins of Newton's First Law, Phys. Teach. 53, 80 (2015)

Hewlett, N., & Beck, J.M.: An Introduction to the Science of Phonetics (1st ed.), Routledge (2013)

Hookway, C.: Peirce, Routledge & Kegan Paul, London, UK (1985)

Kalat, J.W.: Biological psychology, Wadsworth/Thomson Learning, Belmont, CA (2004)

Kamalloo, E., Dziri, N., Clarke, C., Rafiei, D.: Evaluating Open-Domain Question Answering in the Era of Large Language Models, Proc. of the 61st Annual Meeting of the Association for Computational Linguistics, vol. 1, Toronto, Canada (2023)

Kutner, M.L.: Astronomy: A Physical Perspective," Cambridge University Press, pg. 431 (2003)

Van Lieshout, L.L.F., de Lange, F.P., & Cools, R.: Why so curious? Quantifying mechanisms of information seeking Current Opinion in Behavioural Sciences, 35, pp. 112-117 (2020)

Marois, R., Ivanof, J.: Capacity limits of information processing in the brain, TRENDS in Cognitive Sciences vol. 9, no. 6 (2005)

McGowan, J.F.: Kepler's New Astronomy (2009) https://mathblog.com/wp-content/uploads/200 9/12/Keplers-New-Astronomy.pdf

Miller, G.A.: The magical number seven, plus or minus two, The Psychological Review, vol. 63., no. 2, pp. 81-97 (1956)

Newton, I.: The Principia, University of California Press, Berkeley, CA (1999/1687)

Nieder, A., Freedman, D.J., Miller, E.K.: Representation of the quantity of visual items in the primate prefrontal cortex, Science, vol. 297, pp. 1708-1711 (2002)

Nikoo, M.S. et al.: Electrical control of glass-like dynamics in vanadium dioxide for data storage and processing, Nature Electronics 5(9), pp. 596-603 (2022)

Ocklenburg, S., Friedrich, P., Christoph, Schlüter, Beste, C., Güntürkün, O., Genç, E.: Neurite architecture of the planum temporale predicts neurophysiological processing of auditory speech. Science Advances, 4(7) (2018) www.sciencedaily.com/releases/2018/07/180712100455.htm

Odom, G., van de Craats, J.: Elementary problem 3007, American Monthly, 90, p. 482 (1983), solution, 93, p. 572 (1986)

Pascal, B.: Thoughts. In M. Kaufmann (ed.), Cambridge University Press (2013)

Peirce, C.S.: The Essential Peirce: Selected Philosophical Writings, Two volumes, Peirce Edition Project, Indiana University Press, Bloomington (1992, 1998)

Peirce, C.S.: Collected Papers, Harvard University Press, Cambridge, MA (1932)

Peirce, C.S.: A Guess at the Riddle, p. 246 (1887) https://arisbe.sitehost.iu.edu/menu/library/bycsp/ guess/guess.htm

Peirce, C.S.: Logic Notebook, Unpublished manuscript on microfilm, Houghton Library, Harvard University, MA (1865–1909)

Perolat, J., Hennes, D., Tarrasow, E.: Mastering the game of Stratego with model-free multiagent reinforcement learning, Science, vol 378, issue 6623, pp. 990-996 (2022)

Ramachandran, V. S., & Hirstein, W.: Three laws of qualia: What neurology tells us about the biological functions of consciousness. Journal of Consciousness Studies, 4(5-6), pp. 429–457 (1997)

Ries, S.K., Dronkers, N.F., Knight, R.T.: Choosing words: left hemisphere, right hemisphere, or both? Perspective on the lateralization of word retrieval, Annals of the New York Academy of Sciences, vol. 1369, issue 1, pp. 111-131 (2016) https://nyaspubs.onlinelibrary.wiley.com/doi: https://doi.org/10.1111/nyas.12993

Ritter, H., Kohonen, T.: Self-organizing semantic maps. Biological Cybernetics, 61(4), pp. 241-254 (1989)

Rosensohn, William, The Phenomenology of Charles S. Peirce: From the Doctrine of Categories to Phaneroscopy. Amsterdam, B.R. Gruner B.V. (1974) https://en.wikipedia.org/wiki/Phaneron

Sarbo, J.J., Cozijn, R.: Belief in reasoning, Cognitive Systems Research, vol. 55, pp. 245-256 (2019)

Sarbo, J.J., Farkas, J.I., van Breemen, A.J.J.: Knowledge in Formation: A Computational Theory of Interpretation, Berlin, Springer (2011). doi:https://doi.org/10.1007/978-3-642-17089-8

Searle, J.: The Rediscovery of the Mind, MIT Press, Cambridge, MA (1992)

Searle, J.R.: The Problem of Consciousness, Consciousness and Cognition, vol. 2, issue 4, pp. 310-319 (1993)

Seuren, P.A.M.: Western Linguistics: An Historical Introduction, Blackwell Publishers, Ltd. (1998)

Shannon, C.E.: A mathematical theory of communication, Bell System Technical Journal, vol. 27, pp. 379-423 and pp. 623-656 (1948)

Short, T.L.: Peirce's theory of signs, Cambridge University Press, Cambridge, UK (2007)

Stillings, N.A.: Cognitive Science, MIT Press, Cambridge, MA (1998)

Strong, J., Wegstein, J., Tritter, A., Olsztyn, J., Mock, O., Steel, T.: The Problem of Programming Communication with Changing Machines: A Proposed Solution, Communications of the ACM. 1(8), pp. 12–18 (1958)

Taborsky, E.: Energy Transformation and Semiosis. In: M. Bergman & J. Queiroz (eds.), The Commens Encyclopedia: The Digital Encyclopedia of Peirce Studies. New Edition (2003). http://www.commens.org/encyclopedia/article/taborsky-edwina-energy-transformation-and-semiosis

Tee, J., Taylor, D.P.: Is information in the brain represented in continuous or discrete form? IEEE Transactions on Molecular, Biological and Multi-Scale Communications, 6(3), pp. 199-209 (2020)

Weaver, W.: Recent contributions to the mathematical theory of communication. In: C. Shannon & W. Weaver (eds.), The Mathematical Theory of Communication, Urbana University of Illinois Press, pp. 93-117 (1949)

Wittlinger, M., Wehner, R., Wolf, H.: The Ant odometer: Stepping on Stilts and Stumps, Science, vol. 312, pp. 1965-1967 (2006)

Zou, J., He, S., & Zhang, P.: Binocular rivalry from invisible patterns, Proc. of the National Academy of Sciences 113, 30 (2020)

Yang, W., et al.: Selection of experience for memory by hippocampal sharp wave ripples, Science, vol. 383, issue 6690 (2024)

Yarbus, A.L: Eye Movements and Vision, Springer (1967)